ISLAND LIFE

A Natural History of Pulau Tengah, Johor, Malaysia

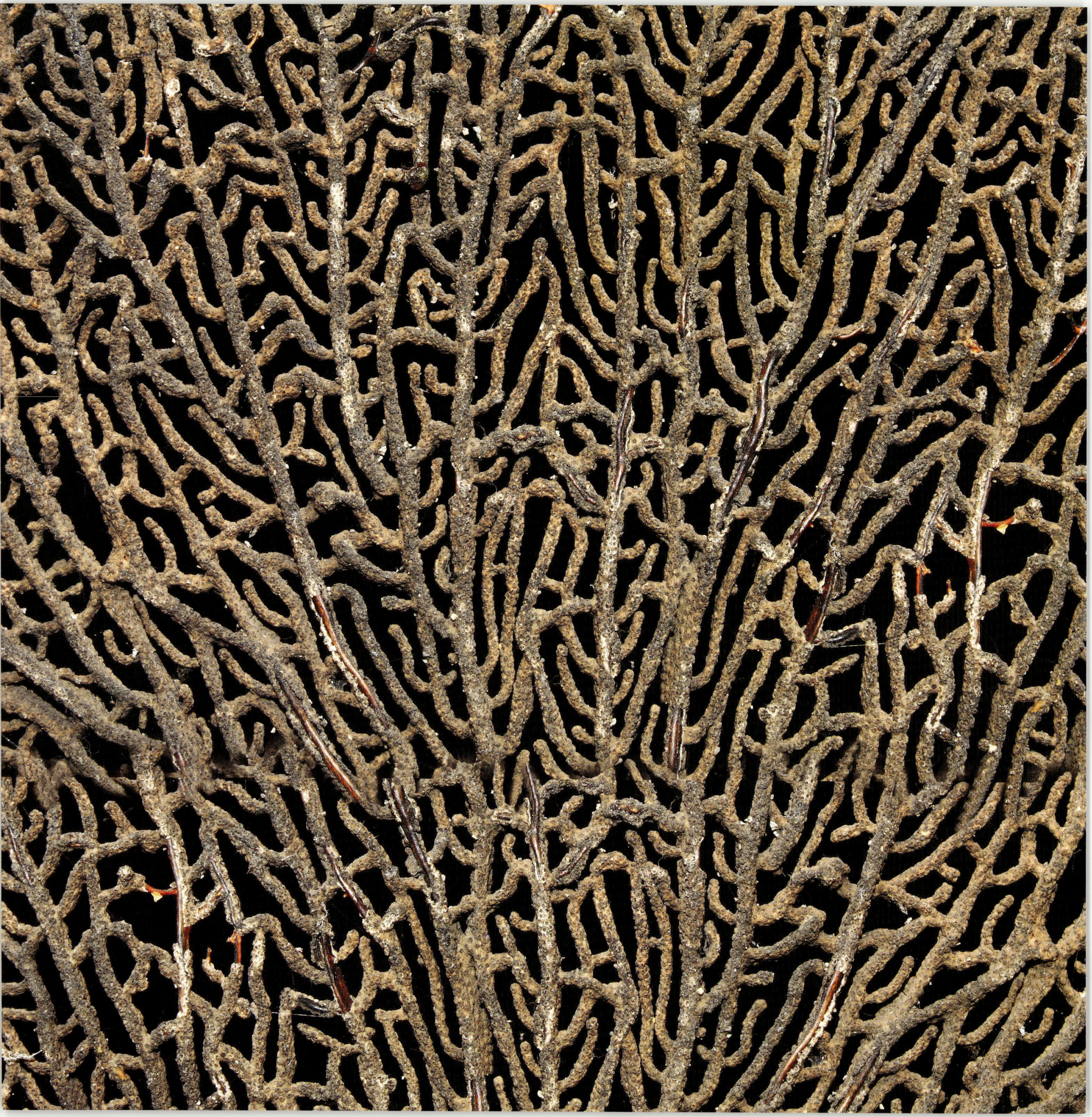

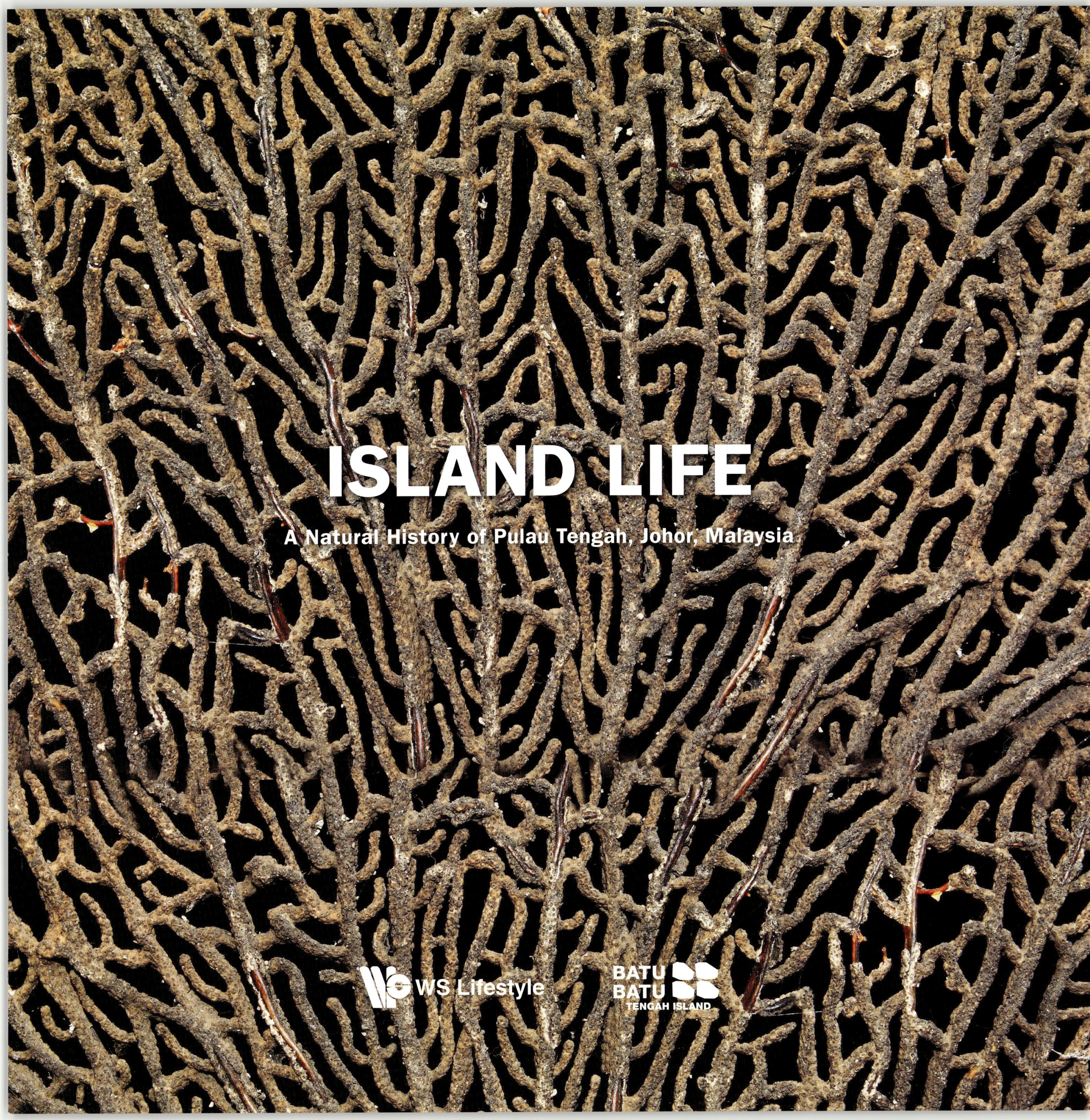

ISLAND LIFE

A Natural History of Pulau Tengah, Johor, Malaysia

WS Lifestyle

BATU BATU
TENGAH ISLAND

Top left: HH Tunku Abdul Jalil Top right: With his father His Majesty Sultan Ibrahim Ibni Almarhum Sultan Iskandar, The Sultan of Johor
Bottom left: At Hyde Park, London Bottom right: With his mother Her Majesty Raja Zarith Sofia binti Almarhum Sultan Idris Shah, The Permaisuri of Johor

This book is dedicated to the late

HH Tunku Abdul Jalil Iskandar ibni Sultan Ibrahim,
Tunku Laksamana Johor

A true lover of nature

Both pages *(Photo credits: The Royal Family of Johor)*

Western flank, Pulau Tengah *(Photo credit: Erik Fearn)*

South-west view of Pulau Tengah *(Photo credit: Erik Fearn)*

Published by

WS Lifestyle, an imprint of
World Scientific Publishing Co. Pte. Ltd.
5 Toh Tuck Link, Singapore 596224
USA office: 27 Warren Street, Suite 401-402, Hackensack, NJ 07601
UK office: 57 Shelton Street, Covent Garden, London WC2H 9HE

National Library Board, Singapore Cataloguing-in-Publication Data

Names: Batu Batu Resort Sdn. Bhd.
Title: Island life : a natural history of Pulau Tengah, Johor, Malaysia / Batu Batu, Tengah Island.
Description: Singapore : World Scientific Publishing Co. Pte. Ltd.,[2015]
Identifiers: OCN 927062318 | ISBN 978-981-4725-45-3 (hardcover)
Subjects: LCSH: Natural history—Malaysia--Johor.
Classification: LCC QH185 | DDC 508.595 --dc23

British Library Cataloguing-in-Publication Data
A catalogue record for this book is available from the British Library.

Cover image: Hawksbill hatchling heading to sea
(Photo credit: Rohan Rees)

ISBN 978-981-4725-45-3

Printed in Singapore

CONTENTS

FOREWORD BY
HIS MAJESTY SULTAN IBRAHIM OF JOHOR

The state of Johor with the longest coastline in Peninsular Malaysia at 400km is bordered on three sides by bodies of water, the Straits of Malacca in the west, the South China Sea in the east, and the Straits of Johor in the south. Johor is gifted with numerous natural wonder including the equatorial forests of Endau-Rompin, the Ramsar mangroves of Kukup and Tanjung Piai, and the Sultan Iskandar Marine Park, home to several endangered and critically endangered species listed under the International Union for conservation of Natura (IUCN). These natural wonders have in turn been declared and demarcated through legislation as nature reserves or protected areas thus allowing for their preservation.

It gives me great pleasure touring my State to meet my people and alleviate their wellbeing and other quality of life. At the same time I advocate not only the cleanliness and the preservation of the environment but also the continued enhancement of Johor's natural riches. The guardianship of Johor's natural assets is a great responsibility that can only be achieved through unity amongst her people. On the one hand, the preservation of Johor's environment and natural richness contributes to the quality of life of her population. On the other, the action and practices of the people have an impact on the natural environment of the State.

Human effort, awareness and dedication are to be in tandem with legislation to ensure the preservation. Global warming, climate change and the impact of human kind on the natural world are the challenges in the conservation of nature if we are to ensure a better quality of life in the future. In this aspect basic education on cleanliness and love of nature is to be initiated among the students and the younger generation.

This book "Island Life — A Natural History of Pulau Tengah" documents just a snapshot of the beauty of the Sultan Iskandar Marine Park and highlights the enormous role of the authorities and the people of Johor have to play. I hope this book will inspire individuals and organizations to join together and focus on efforts in preserving the natural resources and beauty of Johor. This book will spread the beauty of the 87 islands of Johor.

I thank the writers for honouring and dedicating this book to my late beloved son Tunku Abdul Jalil Tunku Laksmana of Johor who was the true lover of nature. I fervently hope that this memory will continue to inspire more conservation work in Johor and beyond.

Thank you.

Yours sincerely,
His Majesty
Sultan Ibrahim Ibni Almarhum Sultan Iskandar
Sultan and Sovereign Ruler of the State and
Territories of Johor Darul Ta'zim
25th April 2017

Peacock Tail Shrimp *(Photo credit: Adam Beard)*

MESSAGE BY MENTERI BESAR

Johor is home to a rich marine life, mesmerizing coral reefs and spectacular sceneries.

Johor is also blessed with an abundance of valuable underwater treasures with 13 pristine islands within its Marine Park in the South China Sea. These immaculate treasures need to be protected and conserved for the future generations to experience.

There is still much in the sea to rejoice, enjoy and protect. And we must never underestimate nature's extraordinary ability to regenerate, given the right circumstances and environment. What remains is ever more precious as species after species teeter on the brink of extinction. We must act now to protect and restore our marine ecosystems, so that they once again teem with abundant wildlife and are able to survive and withstand whatever challenges the future holds.

In Johor, there is no plan B for conservation. We firmly commit ourselves to the principle of sustainability — that we must limit our use of resources to a rate that will serve us not just in the short term, but also for our future generations to enjoy.

Hence, that is why, it is very important to be extra careful and sensitive in carrying out any development on the islands due to the tenuousness and the fragility of the eco-habitat inherent within these islands. The Johor State Government aspires to significantly develop its ecotourism industry in a responsible manner without jeopardizing the environment in line with what the team at Batu Batu has achieved.

The attractions of Johor Marine Parks are actually the abundance of natural resources themselves. Johor has tremendous potential to become a destination for ecotourism should its ecotourism areas be exploited to the maximum. The coral reefs, fish, mangroves, coastal vegetation, beautiful beaches with clear blue waters, and the peaceful and harmonious condition of the islands are some of the enticement of Johor's Marine Park. The combination of all these marine resources has become the main attraction among avid tourists and divers from all around the world. Therefore, it becomes significantly important for us to continue efforts to protect and conserve our Marine Parks.

This means safeguarding and enhancing biodiversity and ecosystems through conservation and stewardship actions, including working on landscapes and seascapes. We must restore Johor's ecosystems. Restoration of degraded ecosystems is our priority, which, once restored, will provide habitat for wildlife and clean water, and are essential for the protection and recovery of species at risk. And we must connect people to nature. Fostering an appreciation for nature and building a "community of stewards" among people of all ages is crucial if sustainability agenda is to be effective.

The Turtle Nest Protection and Hatchery Program in collaboration with the Marine Park Department is a promising initiative to protect and conserve the critically endangered turtle species. This program is highly applauded and I firmly believe that it should be expanded further to other islands in safeguarding these endangered creatures.

This publication is an accolade of treasure trove of scientific information and a showcase of the Johor islands and its Marine Park. I recommend this publication to anyone who has yet to discover the joys of our islands and to rediscover and marvel at the myths, mysteries, beauty, and majestic glory that make up these islands. This is a manifestation of the enthusiasm that the Batu Batu Team has played in taking on marine conservation and sustainable development in the country.

YAB Dato' Mohamed Khaled Bin Nordin
Chief Minister of Johor

Shalling fish being hunted from below. *(Photo credit: Samuel Rohe)*

MESSAGE BY DIRECTOR GENERAL,
Department of Marine Park Malaysia

Malaysia is blessed with rich biodiversity and our islands are home to some of the world's most amazing marine and terrestrial biodiversity. Hence, it is important to document our rich biodiversity not only for scientific purpose but also for the general audience to appreciate this wonderful natural heritage.

This publication, "Island Life — A Natural History of Pulau Tengah, Johor, Malaysia" a coffee table book, is a commendable effort to highlight Babi Tengah Island's rich biodiversity to nature lovers and the general public. This is in line with the global biodiversity targets to enhance awareness on biodiversity at all levels of society by the year 2020. Education and increase in awareness among the public is an important element in ensuring changes in attitudes of the public towards environment and conservation. We are entrusted to safeguard this fragile and sensitive planet for the well-being of all living and non-living resources. We have to be more responsible towards our actions. To ensure successes of conserving our resources, each and every one of us needs to stand on common ground and with common aspirations of sustaining our resources. This has been envisaged in the National Policy on Biological Diversity (1998) and also the Common Vision on Biodiversity (2009).

Records show that Malaysian water is home 77% of total world known hard coral species which support more than 700 reef dependant fish species. Economic value of these areas is estimated to range between RM 10.1 million to RM 3.4 billion base on ecosystem services they provide. As an authority and manager of 42 Marine Parks in Malaysia including Babi Tengah Island Marine Park, the Department of Marine Park Malaysia is committed towards conserving and ensuring the unique marine biological heritage of the nation for the benefit of present and future generations. Therefore, it is with great pleasure that this book has manage to portray the diversity of both marine and terrestrial resources of Babi Tengah Island for the indulgence of visitors to this island.

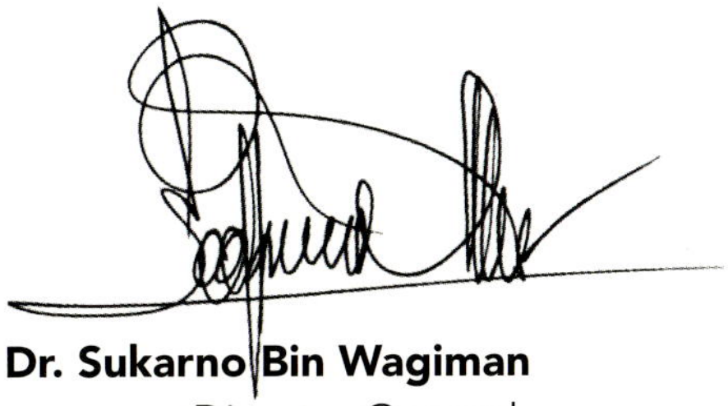

Dr. Sukarno Bin Wagiman
Director General,
Department of Marine Park Malaysia

Rufus, Batu Batu's resident Green turtle. *(Photo credit: Simon Buckell)*

PREFACE

I thank His Majesty Sultan Ibrahim ibni Almarhum Sultan Iskandar for his wise and enlightened Foreward to this book. This book is dedicated to HH the late Tunku Abdul Jalil ibni Sultan Ibrahim for his widely known loving and gentle nature towards all and his care and compassion for wildlife. His Majesty has graciously consented to the naming of our Conservation Centre as The Tunku Abdul Jalil Conservation Centre (TAJCC).

I first set eyes on Pulau Tengah in 1993 when I travelled to meet Pino and Camelia who at the time owned and managed Pirate Bay, a backpacker resort on Pulau Tengah. I fell in love with the island, its white powdered sands, turquoise translucent waters, lush tropical forests — a picture postcard paradise island with its history as a UNHCR Refugee Transit Camp for the displaced Vietnamese Boat people within Johor's pristine Marine Park.

I took an interest in the island in 1994 and for the next 15 years enjoyed it with family and friends — exploring, discovering and enjoying this idyllic paradise. Being a Johorean, however, it seemed a shame that these world class islands should remain unknown to the world and hence the vision of Batu Batu was born. Mindful of the fragile ecology of this small island, it was incumbent upon us, as the sole custodian, to safeguard and protect the natural beauty and rich biodiversity of the island and to nurture the symbiosis between nature and development.

Hence, the architectural intent was to tread as lightly as possible and in a manner that would be harmonious with the natural environment. The development would showcase a Malay Village of "kampung" or village houses made of light timber structures crafted by local carpenters.

Over the last five years, we have started a journey to understand, protect and preserve the natural beauty of Pulau Tengah and its surrounds. We are still at the beginning of this journey but the possibilities to make a difference appear to be endless. This book chronicles and summarises the initial years of work conducted by Nature@Batu Batu in researching, documenting and understanding the history, flora and fauna of Pulau Tengah and in turn developing the appropriate projects to conserve it.

In 2013, Welment A. van Aardenne stayed with us at Batu Batu. She worked for the UNHCR as a human rights lawyer and was posted to Mersing in 1979. As part of her role, she frequently visited the Pulau Tengah transit camp and thus remembers the island vividly when it was largely deforested, filled with makeshift shelters and bursting at its seams with displaced families waiting to be moved onto their new countries. At the end of her stay with us she said, "I am so happy that Pulau Tengah is like this now. It is a happy place." This humbled me with a sense of accomplishment.

I hope that other travellers who set foot on Pulau Tengah will continue to enjoy its exquisite fauna and flora and bring home happy and wonderful memories of our little island.

Dato Chua Jui Leng
Director
Batu Batu, Pulau Tengah

INTRODUCTION

"The real voyage of discovery consists, not in seeking new landscapes, but in having new eyes"

Marcel Proust

The Batu Batu adventure and the journey this enchanting island is taking us on is proof of nature's ability to inspire.

Throughout my childhood we visited the beaches of Johor and snorkelled with big fish over colourful coral reefs. In 1994, my family took an interest in Pulau Tengah and we holidayed on this untouched paradise in its most raw form. After years of scheming and planning our hypothetical dream island resort, my father and I took the plunge in 2009 when the experimental pine forest which he had acquired in the 1970s was ready to harvest. This coincided with my husband and I getting married, and organising a post-wedding, quasi-camping trip on Pulau Tengah with family and our closest friends. It was then, despite intermittent electricity, sandflies, and the least glamorous of circumstances, that we decided to commit to the vision of bringing what would eventually become Batu Batu to life. Soon after, I left my City job in London to return home and embark on this incredible journey.

Having known and enjoyed Pulau Tengah in its natural glory, we wanted to share this secret gem with the rest of the world, while at the same time being careful not to spoil the unspoilt. The plan was therefore to build a low-density tourist destination which had as little impact on the "naturalness" of the island as possible.

From the outset, we put in place the basics to protect Pulau Tengah: we built water treatment plants to ensure we would not have to discharge dirty waters into the ocean; we established procedures to ensure that our boats would not need to throw anchor on the coral reefs or island's seagrass beds; we remained vigilant to ensure that we would not allow our human presence to deter turtles from landing and laying their precious eggs.

When I moved onto the island with our first staff team in March 2012 and opened our doors to our first guests, we started observing, monitoring and recording the island's natural rhythms. We set-up turtle conservation initiatives in 2013 and progressed to create Turtle Watch Camp in 2015. We have had various scientists — ornithologists, botanists, marine biologists — stay with us over time, documenting, photographing and recording the natural life of the island. "Island Life" is an output of these early years and a culmination of this initial period of basic protections and observations — in my mind, a marker for the end of "Phase One".

And Phase Two? I believe Phase Two stems from the key lesson learned in Phase One — that we not only have the ability and responsibility to protect the island (an ongoing endeavour and something which will I suspect be forever work-in-progress), but we also have the power and privilege to give back to nature. In fact, the coral and fish populations around the island appear better today than they were before we opened Batu Batu. By protecting our reefs and letting them grow healthier day-by-day, by enforcing the no-take zone around the island, by protecting turtle eggs from poachers which allows thousands of endangered green turtles and critically endangered hawksbill turtles to hatch, by picking-up hundreds of bags of washed-up plastic from our beaches each year — in all of these ways, we can have a positive impact on the island and her surrounds.

So, in this day where half of the world's coral reefs have been lost in the last 30 years and climate change is a horrifying reality, where we read in our newspapers that 90% of all coral will be extinct by 2050 and we wonder if our grandchildren will ever snorkel with Nemo, we are taking up the massive challenge of trying to create a sustainable island community model. A tourism business which tries its very best to minimise negative environmental impact while running conservation initiatives and changing the mindsets of our staff, our guests and the surrounding communities. And by embarking on this journey, we hope to show others that this environmentally sustainable model for a tourism business can be an economic positive in the hope that some will follow suit.

We are fortunate to have the support of guests whose genuine eagerness and interest in conservation and sustainability encourage us to continue pursuing our passion. In fact, my seven-year-old son's favourite activity on the island is the weekly beach clean-up run by our Turtle Watch Camp.

We know that we have not picked the easiest option and we will undoubtedly face numerous struggles and frustrations along the way. Will we succeed? I don't know, but I strongly believe it is important that we try — try to push boundaries and try to create change.

We have already started taking our first steps in Phase Two, defining our aims and exploring ways for us to best achieve our goals. In doing this we have heard the most incredible stories and met the most amazing and inspirational people: the farmers turning raw food waste into compost and biogas, creating the most fertile soils for growing organic fruit and vegetables, and the entrepreneurs bringing sustainable solutions to the forefront, saving precious water, reducing the use of fossil fuels. We look forward to recounting our experiences and encounters on our Batu Batu blog as we move along.

I sincerely hope that we can inspire others and together make the changes that will allow our children and grandchildren to happily explore and continue to be amazed by the shimmering coral reefs around these magical islands.

Wish us luck!

Cher Chua-Lassalvy
Co-owner & Managing Director
Batu Batu – Tengah Island
November 2017

Long Beach (*Photo credit: Erik Fearn*)

PULAU TENGAH

Latitude – 2°28’37.02”N
Longitude –103°57’38.49”E

Nine nautical miles off the east coast of peninsular Malaysia, strung along the middle arc of the Seribuat Archipelago, the roving eye will spot a clutch of sun-washed islands.

Narrow the lens a little to focus on the middle island, Pulau Babi Tengah (or Middle Pig Island in the Malay language) named after the wild pigs that used to roam its lands.

Set in the protected Johor Marine Park, the island, better known by its shortened name of Pulau Tengah, is three kilometres in circumference with an elevation of 150m at its highest point. Its coastal fringes feature fine powder white sandy beaches interspersed with rock outcrops.

The climate of the island is typically tropical. Temperatures at night rarely drop below 24 degrees Celcius and hover around 30 degrees by day. From March through to November, the south westerly winds blow and the season is characterised by sunny, blue-skied days and calmer seas. From December through to February, the north easterly winds bring with it a higher chance of rough seas and rain.

Though just 16 kilometres from the fishing town of Mersing, Johor, and 140km from Singapore, the island ticks every fantasy of an uninhabited paradise island.

The beaches that encircle most of the island attract Green and Hawksbill turtles that land to lay their eggs from March to October. The translucent waters that surround the island are home to both coral reefs and meadows of sea grass which in themselves house rich marine life.

The island has not always been completely uninhabited. In the late 1970s, Pulau Tengah was home to over 100,000 Vietnamese refugees when it was a UNHCR Refugee Camp for the Vietnamese Boat People waiting to be re-housed in new countries. In later years, the island was home to a small backpacker resort called Pirate Bay and more recently it served as a set for Expedition Robinson and Koh Lanta — two European survival reality television series.

Batu Batu sits on the southern end of the island and is built in the traditional Malaysian "kampung" or village style to blend into the natural landscape of the island. The resort aims to tread lightly in order to preserve the beauty of the island and its natural surroundings.

To this end, Batu Batu has set up a variety of projects and funded a number of studies over the past years, including an ongoing collaboration with Malaysia's National Marine Parks Department for the conservation of turtles.

This book sets out to document and share a broad overview of the natural history of Pulau Tengah. It is a dedication to the work of Batu Batu's staff and the various experts and nature lovers who have visited the island and contributed a little to its preservation.

Photo credit: Jabatan Taman Laut Malaysia. Kementerian Sumber Asli dan Alam Sekitar

The Johor Marine Park

Under the Fisheries Act 1985 (amended 1993), 42 islands and their surrounding waters off the east coast of peninsular Malaysia in the South China Sea are gazetted as the Malaysian Marine Park. The introduction of this legislation was both a recognition of the rich biodiversity of the area's marine flora and fauna; and a desire to protect and conserve their natural habitat.

The Johor Marine park consists of 13 islands in a sea-zoned area for a distance of two nautical miles from the low tide line of the islands. Pulau Babi Tengah sits within a cluster of seven islands namely: Pulau Babi Tengah, Pulau Babi Besar, Babi Hujung, Pulau Rawa, Pulau Gual, Pulau Mensirip and Pulau Harimau.

The other cluster of six is toward the south and east of Pulau Babi Tengah: Pulau Tinggi, Mertinggi, Sibu, Sibu Hujong, Pemmanggil and Aur. The topography of these islands is characterized by narrow coastal fringes, rising sharply to rocky ridges and escarpments.

MARINE LIFE

To the landlocked eye, Pulau Tengah's landscape is one dominated by tropical forests, fine sandy beaches, clear waters and rocky outcrops. However, a second world of incomparable beauty and rich biodiversity lies, literally, just below the surface of the sea.

Indeed, the marine environment of Pulau Tengah, Mersing and its neighbouring islands is diverse enough that various species of plant life can thrive, while a vast flotilla of marine mammals are able to call it home. Protected as national and state marine parks, the blue and emerald waters are a haven for coral reefs, as well as playgrounds and nesting points for the dugong and at least 26 species of whales, dolphins and porpoises.

The creeping reach of human activity — not least poaching, commercial fishing, dredging and coastal development — has meant that this delicate environment is under constant threat. As sentinels of the ocean's health, marine mammals, seagrass and coral reefs play a critical role in the preservation of a healthy marine environment. In turn, this has important consequences for humanity's long-term survival which we ignore at our peril.

Left: Clownfish *(Photo credit: Adam Beard)*

Top: A bed of Acropora sp. colonies.
Left: White eyed moray eel in coral garden. *(Photo credits: Adam Beard)*

Top: Green turtle nesting at Turtle Beach with turtle volunteer Steven.

THE TURTLES OF PULAU TENGAH

The turtle has lived in our oceans for over 150 million years, one of the few species to have outlasted the dinosaurs. Larger than its land relatives, it cannot retract its head completely into its compressed stream-lined shell. Flattened forelimbs act as underwater propellers while smaller hind legs help to steer. Though technically an air breathing animal, a turtle is able to spend large amounts of time underwater, coming up for air approximately once every twenty to thirty minutes when awake.

Today, turtle numbers have diminished significantly — a reduction caused in large part by human activity such as trawling and the trade in turtle meats, shells and eggs.

Only seven marine turtle species remain worldwide — the Leatherback, Green, Loggerhead, Kemp's Ridley, Olive Ridley, Hawksbill and the Flatback. Of these seven, four can be found in Malaysia: the Leatherback, the Green, the Hawksbill and the Olive Ridley.

Though efforts to protect Malaysia's turtles date as far back as the 1950s, their survival continues to be threatened. The Leatherback, once a common enough visitor to the area, has not been seen in years, while the Green is listed as endangered and the Hawksbill as critically endangered under the IUCN Red List.

Pulau Tengah is fortunate to be the nesting ground for the Green and Hawksbill turtles. Indeed, the turtles are some of the island's most beloved residents, so much so that their conservation lies at the heart of Batu Batu's conservation programme. For several months in 2012, our opening year, a huge completely tame Green Turtle nicknamed Rufus could often be found paddling around the resort and its waters.

Green turtle nesting at the Turtle Beach.

Freshly laid turtle eggs. *(Photo credits: Simon Buckell)*

Every year, mainly between March and October, female Green and Hawksbills turtles come ashore — usually under the cover of night — to lay their eggs on one of the over ten beaches on Pulau Tengah. Sometimes, they come from thousands of miles away to lay their eggs on the same beach on which they had hatched some three decades ago.

The female turtle crawls up the beach to a dry area above the high tide line where she proceeds to flip away loose sand with her flippers. She then constructs a body pit by digging with her flippers while rotating her body. Using her hind flippers like a shovel, she digs a cavity into which she proceeds to lay her eggs.

During this process, the mother appears to shed tears, but she is actually secreting salt that accumulates in her body. The eggs are flexible and coated with a layer of mucus which allows for up to 200 eggs to be laid at a time. After an hour or so, she uses her rear flippers to cover the pit and throws sand in all directions to throw off predators.

Her maternal task complete, she heads back into the sea and never returns, not even to tend the nest. However, she returns to the same area ten to fifteen days later to lay again. It is at this stage that the nest is most vulnerable to predators such as monitor lizards, crabs and snakes; human poachers dig up the eggs and sell them for up to three Malaysian Ringgit each.

Should the nest remain undisturbed, the incubation period is usually between 50 and 70 days. Not only do warmer temperatures shorten the incubation period, they also tend to produce females; cooler temperatures produce males.

Baby turtles usually hatch in the cool of the night or during a thunderstorm in the day. The entire nest hatches at the same time and the baby turtles all scramble into the sea together, attracted by the moon's reflection on the horizon. Here, the hatchlings face many predators, both on land (crabs and birds) and in the sea (fish).

But as the hatchlings head toward the sea, they leave a magnetic imprint on the sand which enables them to return decades later to nest in the same spot where they were hatched. It is theorised that turtles can detect the angle and intensity of the earth's magnetic field and hence the latitude and longitude of their position. This remarkable ability enables the turtle to navigate anywhere in the oceans.

Turtle Conservation

Turtles face many obstacles to their survival as a species.

Many become entangled in fishing nets. Their natural habitat is littered with man-made debris. Autopsies of many turtles reveal plastic bags in their stomachs — inadvertently swallowed because in the water, the bags resemble jellyfish, their favourite food. Coastal degradation, relentless poaching, and the loss of nesting grounds complete the grim tale of slow extinction.

As we have seen, their eggs and hatchlings are vulnerable to predators and human poachers.

Working with the Johor Marine Park and the Department of Fisheries, Batu Batu's conservation team "Turtle Watch Camp" has diligently documented landing and nesting patterns. The ideal has been to allow the eggs to incubate *in situ*, ie. where they have been laid. However, as the team are unable to guard the nests from poachers and monitor lizards around the clock, the decision was taken to set-up a fenced-in turtle hatchery at Long Beach in the Spring of 2014. The egg chamber is carefully dug to resemble as much as possible their natural nest, and then fenced off against human poachers, lizards and rats.

Between April 2014 and October 2017, more than 20,000 eggs were located to the hatchery. Of this number, 53% hatched and the hatchlings were successfully released into the sea. Approximately 80% of the hatchlings were Hawksbill and 20% Green turtles. At all times, the conservation team has followed strict handling and release protocols.

Turtle hatchery at Long Beach.

Batu Batu construction manager Azmi (left) and Kevin Gryce (right) at the hatchery.

Photo credit: Sam Crouch

Turtle hatchery at Long Beach. *(Photo credit: Rohan Rees)*

Batu Batu guests watching the release of Hawksbill hatchlings. *(Photo credit: Adam Beard)*

Hatchlings swimming into the deep sea. *(Photo credit: Simon Buckell)*

Beach tracks made by nesting turtle.

Nest disturbed by poachers. *(Both photo credits: Simon Buckell)*

(Photo credit: Simon Buckell)

The Green Turtle

Weighing up to 350 pounds, the large Green turtle boasts a carapace that is over a metre long, and features a kaleidoscope of colours including black, grey, green, brown and yellow. Immature Greens are carnivorous but, on achieving adulthood, they switch to a herbivorous diet of seaweed and seagrass — found in abundance off the beaches of Batu Batu — an odd trait not shared by any other turtle species. An interesting consequence of this diet is that its fat is literally green in hue, hence its name.

(Photo credit: Shutterstock)

The Hawksbill Turtle

Named for the raptor-like shape of its beak, the Hawksbill turtle has been hunted nearly to extinction for the beautifully patterned carapace. Growing to a metre in length and weighing as much as 180 pounds, it feeds mainly on sponges and seagrass. Happily for the Hawksbill, the coral reef within the Mersing Marine Park hosts a symbiotic eco-system of corals, invertebrates like mollusks and anemones, algae, seagrass and seaweeds.

CORAL

A healthy coral reef ecosystem is comprised of three main groups: coral, coral reef invertebrates, and reef fish.

There are two different types of coral; hard and soft. Hard corals secrete a calcium carbonate exoskeleton that forms the reef structures often seen. Each coral structure is comprised of a colony of individual coral polyps.

Corals have a symbiotic relationship with microscopic algae, zooxanthellae. The coral provides shelter and nutrients for the algae, and in return the algae shares the energy it produces through photosynthesis. It is thought that the zooxanthellae provides over half the energy needed by the corals. The zooxanthellae are also what give the corals all of their magnificent colours.

Coral is part of the phylum Cnidaria. Like anemones and jelly, to which they are related, coral polyps are predators. Their tentacles are full of stinging cells that they use to kill any prey that happens to float by.

There are thousands of coral species in the world. Malaysia alone boasts around 500 to 600 species. Each coral species features individual, unique growth forms. Collectively, they create the complex architectural reef structure upon which so many marine life-forms depend.

With some of the highest biodiversity of marine animals, a coral reef is nature's most productive marine community.

The interactions between the various fauna and flora species in the reef create a complex self-sustaining ecosystem.

It is an important nursery ground for many marine organisms. It plays a crucial role in protecting coast-lines: by acting as barriers and by absorbing wave energy, it shelters islands such as Pulau Tengah.

Many important new pharmaceutical drugs are derived from reef organisms. In fact, the first active compound from marine organisms was found in the sea-fan, or gorgonian coral. Following that discovery, bio-compounds from marine organisms were extracted to help fight cancers.

Some of Pulau Tengah's best coral reefs are located along its northern and southern shores. Individually identified as the Northern Reef, Southern Reef, Junior Reef and Eastern Reefs, this complex necklace of reef ecosystem has a depth range of three to eight metres with diverse coral reef communities. Here, aided by regular circulating currents, the clear and shallow waters provide perfect spawning conditions.

Batu Batu's resident scuba diving instructors and Turtle Watch Camp team have spent long periods in the water documenting the island's coral habitat. Recorded types of coral include branching corals, cauliflower coral, mushroom corals, and boulder coral.

Left: Galaxea sp. (Photo credit: Adam Beard)

Favids, *Montastrea sp. 1.*

Oxypora, *Montastrea sp. 1.*

Acropora sp.2.

Acropora sp.1.

Favids, *Montastrea sp. 2.*

All photo credits: Kee Alfian

Bubble coral, *Montastrea sp. 2.*

Coral consists of polyps with six tentacles that are protected by a calcium carbonate skeleton structure. In this close-up, the polyp is extended to create more surface for sunlight absorbtion (photosynthesis) and feeding on plankton for nutrients.

Featherstars are highly mobile, but most of the time they cling to substrates using a hair-like structure called cirri. (*Photo credits: Adam Beard)*

Crown-of-Thorns play both positive and negative ecological roles in the coral reef but they are mostly known to cause death to a reef with their huge appetite for fast-growing coral. (*Photo credits: Kee Alfian)*

Featherstars sway their feathery arms in the current and capture organic particles and plankton that stick to their pinnules. The food flows down to their mouths through a mucous canal. (*Photo credits: Kee Alfian*)

CORAL REEF INVERTEBRATES

Reef invertebrates are a fascinating sub-group of marine animals that includes the sea star, sea cucumber, sea urchin, sea anemone, mollusc, and crustaceans that live in the coral reef ecosystem. They form a harmonious, mutually beneficial relationship with the coral reef.

Sea Stars

The coral reefs of Pulau Tengah are home to featherstars, and sea stars such as *Fromia monilis* and *Acanthaster planci*. Though the latter — better known as the Crown-of-Thorns — has, in recent years, gained some notoriety for destroying vast swathes of Australia's Great Barrier Reef, it is, in fact, nature's answer to maintaining the delicate balance of a reef's ecosystem. After all, the survival of a thriving reef depends on a multiplicity of corals, all of which grow at different rates. In its proper role, the Crown-of-Thorns prevents a reef from being unbalanced by a speedily growing coral species.

Collector urchin, *Tripneustes sp.*

Banded sea urchin, *Echinotrix calamaris*.

Red urchin, *Parasalenia gratiosa*.

Longspine urchin, *Diadema setosum*.

Sea Urchin

Another Pulau Tengah resident is the sea urchin. A conspicuous presence in the surrounding waters, this marine animal is rightly feared by snorkelers and scuba divers for the pain it inflicts should they accidentally apply pressure from any part of their body onto the delicate but toxic spines. Still, sea urchins, especially their roe, are a prized delicacy in Japanese restaurants. What's more, as the reef's *de facto* gardener, they feed happily on algae that could potentially outgrow coral and turn an entire coral reef to an algae ecosystem.

All photo credits: Kee Alfian

Spiky sea cucumber, *Stichopus chloronotus.*

Pink and black sea cucumber, *Holothuria edulis.*

Cushion star, *Culcita novaguineae.*

Sea Cucumber

The sea cucumber is a close cousin of the sea urchin, though it lacks the poisonous spine. Valued in the dry seafood market, especially in Southeast Asia, they also boast medicinal properties — 'minyak gamat' is a locally made ointment for cuts and bruises. The sea cucumber has other uses, not least its status as a marine vacuum cleaner that feeds on organic debris on the surface of coral and other hard substrates that could potentially create damaging bacteria. Because their numbers in the wild are decreasing, it's been mooted that protected areas such as Pulau Tengah could be used as breeding grounds.

Riterri anemone, *Heteractis magnifica (Photo credit: Adam Beard)*

Tube anemone, *Cerianthus sp. (Photo credit: Kee Alfian)*

Anemone shrimp, *Periclemens sp. (Photo credit: Kee Alfian)*

Riterri anemone, *Heteractis magnifica (Photo credit: Adam Beard)*

Riterri anemone, *Heteractis magnifica* with clownfish.
(Photo credit: Kee Alfian)

Leathery sea anemone, *Heteractis crispa*
(Photo credits: Kevin Gryce & Anne Jeanneret)

Sea Anemone

There are a thousand species of sea anemones, but only ten are known to be host to anemone fish. Of that ten, six are found in Malaysian waters.

One of the marquee attractions of a coral reef, the sea anemone is a pretty but predatory marine animal. Yet, with very little fanfare, it engages in a classic symbiotic relationship with various planktonic plants and organisms such as the anemone shrimp and juvenile butterfly fish.

In particular, its softly undulating tentacles, which hide an assorted arsenal of poisonous defences, play host to a certain group of damselfish, better known as clownfish (better known to the Disney generation as Nemo).

This fish depends on the sea anemone for shelter from predation, protecting itself from the venomous tentacles by a combination of unusual swimming behaviour and the production of special chemicals in its skin. In exchange, the clownfish helps preen the sea anemone's tentacles, keeping it in a healthy condition.

Sea anemones contain many biologically active properties that are important in the medical fields of neurophysiological and pharmacological researches. However, their fate is intricately tied to the fate of its symbiotic partners. Indeed, their very presence provides a reliable indication of ultra-violet (UV) radiation levels in the marine environment. This is important as marine organism larvae are particularly susceptible to UV radiation.

Chromodoris coi (Photo credit: Adam Beard)

Nudibranchia (Photo credit: Adam Beard)

Tridacna squamosa (Photo credit: Kee Alfian)

Tridacna crocea (Photo credit: Kee Alfian)

Mollusc

The mollusc is one of the largest groups of reef invertebrates in the coral reef ecosystem. The majority of these shelled organisms feed by filtering seawater; some prey on fish and other shelled organisms.

The giant clam is among the largest of the species. Though it is protected by Malaysia's Fisheries Act, its shell is often collected as a souvenir while some island natives eat its meat. Sadly, its wild population is decreasing and the biggest of its species, the *Tridacna gigas*, which can grow up to two to three metres is extinct in the waters of Peninsular Malaysia; the waters of Sabah are the only place that this species still thrives.

The nudibranch or the naked shell is another familiar mollusc. Sporting vibrant colours, it is one of the most photographed organisms of the reef. Fish will not feed on it as it is toxic. The odd thing is that the nudibranch does not actually produce these toxins; rather, it accumulates them by feeding on stinging hydroids, sponges and algae on the reef.

The cone shell is another species group of mollusc that preys on fish. It has a harpoon-like tooth that can be extended to deliver toxins that enable it to paralyse and then devour its prey whole. Cone shells, especially those with unique patterns, are prized by human collectors though collecting them can be hazardous. They release a toxin that can paralyse a human's ability to breathe which has led to unfortunate drowning incidents.

Long-beak butterflyfish, *Chelmon rostratus*
(Photo credits: Kevin Gryce & Anne Jeanneret)

Threefin fish, *Helcogramma sp. (Photo credit: Kee Alfian)*

CORAL FISH

Reef fish are an important community of the coral reef ecosystem. Their greater mobility allows them to roam larger areas of the reef.

Herbivore reef fish feed on algae and, together with the reef's herbivorous invertebrates, they help balance the delicate relationship between the coral and algae in the reef. Indeed, the feeding pattern and feeding area of some reef fish such as the butterfly fish are important indicators of the health of a coral reef.

Carnivorous reef fish like the snapper, garouper, sweetlips and barramundi cod are highly prized for human consumption. This fact alone makes protecting the reef habitat critical especially if this group of fish continues to be heavily fished.

Coral fish are also important in maintaining healthy numbers of reef sharks; in turn, the latter are, themselves, indicators of a healthy reef.

The British naturalist Simon Buckell has reported seeing black-tip reef sharks consistently around the coral reefs at Pulau Tengah. These sharks can sometimes be seen near the jetty, or roaming in and out between the reef and the seagrass meadows.

Lionfish, *Pterois sp. (Photo credits: Kevin Gryce & Anne Jeanneret)*

Catfish *(Photo credit: Adam Beard)*

Sweetlips Reef Fish, *Plectorhinchus sp.*
(Photo credits: Kevin Gryce & Anne Jeanneret)

Freckled Goatfish, *Upeneus tragula*
(Photo credits: Kevin Gryce & Anne Jeanneret)

Black-spotted Pufferfish, *Arothron nigropunctatus*
(Photo credits: Kevin Gryce & Anne Jeanneret)

Batfish, *Orbicularis sp. (Photo credit: Adam Beard)*

BLACKTIP REEF SHARK

Juvenile and adult blacktip reef sharks are commonly seen close to the shores of Pulau Tengah. These are friendly, harmless creatures. *(Photo credit: Sam Crouch)*

Chaetodermis penicilligerus (Photo credit: Adam Beard)

SEAGRASS

A thriving seagrass ecosystem can be found to the west of Pulau Tengah and small patches around the island.

Dating back millions of years to the age of the dinosaurs, seagrass are flowering plants that live exclusively in a marine environment. Just like land plants, they conduct photosynthesis and produce oxygen which makes their population size largely dependent on the clarity of the water.

One of the main reasons why Pulau Tengah's seagrass meadows have thrived is that their lack of visual appeal has meant that scuba diving enthusiasts prefer the more eye-popping spectacles of coral reefs. That said, the marine residents including crustaceans have developed remarkable camouflaging abilities especially in the monotone qualities of their skin colouring.

Despite their comparatively low-key appearance, seagrass play an ecological role that is just as important as coral reefs. They are, as we've seen, critical to the dugong as well as being important habitats for seahorses, turtles and fish fry.

In particular, seagrass ecosystems reduce and stabilise sediment loads to coral reefs. Its photosynthesis capacity is staggering.

A single hectare of seagrass produces almost 100,000 litres of oxygen a day while absorbing ten times more carbon dioxide than an equivalent area of rainforest.

The current lack of conservation efforts in this regard is a critical lapse that will have catastrophic consequences.

Top: Dugongs swimming near the water's surface at the jetty of Pulau Tengah. *(Photo credit: Muhammad Sholihuddin)*
Left: Dugong feeding on seagrass. *(Photo credit: Shutterstock)*

THE DUGONG

The dugong is a gentle giant. Shy and reclusive with oversized paddle-like front flippers, this large marine mammal herbivore is, somewhat bizarrely, a close relative of the elephant, sharing many internal and external morphological characteristics of its terrestrial cousin.

Comparatively little is known about the dugong. Called a duyung, or mermaid, by the indigenous Seletar people who live along the coast of the Johor Straits, the creature is endangered in most parts of the world, especially around the Johor waters. (The largest populations are found in Australia and the Arabian Gulf.)

In 1999, it caught national media attention in Malaysia when a dugong calf was caught in a fisherman's net in the eastern Johor Straits. Before this, the public either had no knowledge of the animal, or that it even existed in Malaysia. In the following years, research teams led by local and international marine scientists conducted intensive studies.

The results have been a little discouraging. Dugongs are believed to have once been a common denizen in the Johor Straits, the narrow waterway separating Malaysia and Singapore. Modern development and industrialisation on both sides of the straits have worked their predictable damage. Today, it appears that the Mersing Islands off Johor — with their extensive seagrass meadows — are the last stronghold in Peninsular Malaysia where dugongs can be found with certainty.

A large group of dugongs swimming near the water's surface, over a large area of verdant subtidal seagrass meadow off nearby Pulau Sibu Tengah. *(Photo credit: Dr Louisa Ponnampalam, The MareCet Research Organization)*

Each day, the average adult dugong feeds on an estimated 50kg of seagrass which are rich in nitrogen and starch content. It is for this simple reason that the Mersing Islands, with their verdant underwater meadows of seagrass, are so important for dugongs occupying the area.

Dugongs are most commonly found at Pulau Tinggi and Pulau Sibu Hujung though there have been sporadic sightings near the jetty of Batu Batu, and the channel between Pulau Tengah and Pulau Besar.

Reclusive and retiring, dugongs do not engage in any showy acrobatics. Sightings of the notoriously shy mammal invariably involve a glimpse of either its nostrils, or of its back and tail as it dives after taking a breath at the surface. Underwater excursions may yield sightings of feeding trails on the seagrass meadows, much like those of freshly mowed lawn, which are made by the dugong as it shuffles along uprooting and munching on choice seagrasses.

The dugong is inseparable from its seagrass habitat. These meadows are crucial for maintaining not just the health of the marine environment, but also the physical integrity of the coast which, in turn, has practical long-term repercussions for the human urbanite. More critically, the current research indicates that the most important dugong areas lie outside the legal boundaries and protection of the marine parks.

Yet, because seagrass meadows lack the visual punch of a multi-coloured coral reef, their importance does not translate into the same conservation urge. Beyond the confines of marine parks, seagrasses raise barely a mention in conservation policies, leaving them at the mercy of human activities like dredging, trawling and pollution.

The silver lining is this: dugongs need seagrass to survive, and dugongs have a massive public appeal. This syllogistic synergy makes the dugong the perfect mascot for seagrass conservation which, in turn, bodes well for humanity's own long-term well-being.

For now, the task is to deepen our understanding of dugongs and their ecology. The fact that they seem to have made the accessible waters around Pulau Tengah and Mersing their habitual foraging ground makes the job of marine biologists a little easier. But considerably more work is needed to develop, with key stakeholders, a conservation action plan that covers both dugong and key seagrass meadows.

Right: Dugong feeding on seagrass. *(Photo credit: Shutterstock)*

An Indo-Pacific bottlenose dolphin near Pulau Chebeh, north-east of Pulau Tengah.

An Indo-Pacific bottlenose dolphin looks curiously at the photographer, as it swims in the clear waters of the South China Sea offshore from Pulau Tengah. Its uniformly grey colouration is evident here.

A group of Indo-Pacific humpback dolphins surfacing close to each other. The mottled grey-pink colouration of this species' skin is a distinctive characteristic in adults.

A pair of the usually elusive Indo-Pacific finless porpoise surfaces gregariously. This species is typically difficult to observe and rarely leaps out of the water when surfacing. *(All photo credits: Dr Louisa Ponnampalam, The MareCet Research Organization)*

DOLPHINS AND PORPOISES

The blue and emerald waters of the marine park around the Mersing Islands including Pulau Tengah are also home to the Indo-Pacific humpback dolphin, Indo-Pacific bottlenose dolphin, long-beaked common dolphin, and the Indo-Pacific finless porpoise.

The stumpy Indo-Pacific humpback dolphins prefer the turbid coastal waters, while the shallow areas between the islands and the mainland are favoured by the elusive, finless porpoises. A little further offshore, one is more likely to encounter the more open-sea species of Indo-Pacific bottlenose dolphins and gregarious long-beaked common dolphins.

The Indo-Pacific Humpback Dolphin

Preferring the often turbid coastal waters, the stumpy Indo-Pacific humpback dolphin is a rather peculiar species: it is the only one of its brethren that is born pale grey, before turning almost entirely pastel pink with grey mottling as it grows into adulthood.

It has been seen around Batu Batu near the jetty engaging in dazzling displays of aquatic acrobatics — leaping out of the water and slapping its tail — or swimming parallel with a moving boat, though it's more likely to be found near mangroves, especially the mangrove shorelines along the southern end of Pulau Sibu Tengah.

The Indo-Pacific Finless Porpoise

The only porpoise species in Malaysia derives its name from its lack of a dorsal fin. Often dubbed a "floating tyre" by marine mammal scientists, it measures no more than two metres in length and sports a rounded head with spade-shaped teeth.

This black beauty has a famously retiring nature; it prefers the shallow waters between Pulau Tengah, its neighbouring islands and the mainland, though it has been sighted near the jetty of Batu Batu. Having no dorsal fin, it is difficult to spot in its natural habitat. Indeed, sightings are generally accompanied by incredibly calm weather or sheer luck and even then, it's usual to only get a fleeting glimpse of its back. Its low profile is compounded by the fact that it does not leap out of the water, nor does it engage in showy acrobatics. It's only when it is socialising or feeding that its guard is down, and one is able to see its pectoral fins or tips of its tail flukes bobbing out of the water.

The Indo-Pacific Bottlenose Dolphins and Long Beaked Common Dolphins

A little further offshore in the crystal clear waters, one encounters the more "open sea" species such as Indo-Pacific bottlenose dolphins and gregarious long-beaked common dolphins.

The former — characterised by a largely grey and uniform skin colouration, with a pale and sometimes flushed pink belly and some grey spotting, and an unmistakably tall and curved dorsal fin — is what most people tend to think of as dolphins. This is, in large part, due to their starring roles in Hollywood movies like 'Flipper' and 'A Dolphin's Tale'.

Indeed, the Indo-Pacific bottlenose dolphin is most likely to be the mammal spotted by Batu Batu guests while diving or snorkeling. Frequently sighted around Pulau Tengah and the small rocky island of Tokong Bahara towards the northeast, the species roams the deeper and offshore waters of the area, swimming closer to the islands from time to time to feed on reef fish.

Given its naturally upward-curved mouth-line that causes it to appear as if it is smiling, the bottlenose dolphin is a deceptively friendly creature. It is, in fact, one of the sea's most active predators. It is advisable never to jump into the water to swim with them. They are best watched from a safe distance within the boat.

Tidal pool at Lagoon Bay.

TIDAL POOL

Two interesting features of Pulau Tengah are the tidal pools found on the island's northern beaches and on Sunrise Beach in the south at Batu Batu. Exposed during low tide — when sea water is trapped in depressions in the coral and rock — the phenomenon teems with all manner of life including coral that have adapted to living in extreme environments that are alternately wet and dry, as well as rock clams, oysters, and barnacles. Crabs hide beneath stones, while fish and sea stars lurk amongst seaweed, soft coral and seagrass.

Being so exposed brings its own dangers as the heat and air can desiccate delicate flesh and leaves. Without the protective blanket of water, death is always imminent.

Come high tide and these plants and animals must now somehow cling to the rocks or risk being dislodged by the power of the surging water and swept out to sea or, equally disastrous, tossed onto the beach.

Tidal pools at Red Sands Beach. *(Photo credit: Dave Bakewell)*

Tidal pool at Sunrise Beach.

Corals, as we've already seen, are true marine animals that generally cannot survive outside their habitat. However, the coral in Pulau Tengah's tide pool have adapted to the dry phase by secreting a layer of mucus that covers the whole coral structure and keeps it moist till the tide comes back in.

Indeed, the cyclical nature of the tidal environment means that the animate inhabitants of the tide pool such as crabs, bivalves and sea cucumbers have evolved a special process that allows them to capture their food by filtering seawater or by casting a biological feeding net. Sea cucumbers, for instance, use a very sticky extension in their mouths to catch organic materials and planktons. The intertwining of this extension forms a net-like structure which they retract back into their mouths for feeding.

Red Hermit Crab, *Dardanus megistos*.

Granulated Flatfoot
(very poisonous)
Platypodia granulosa

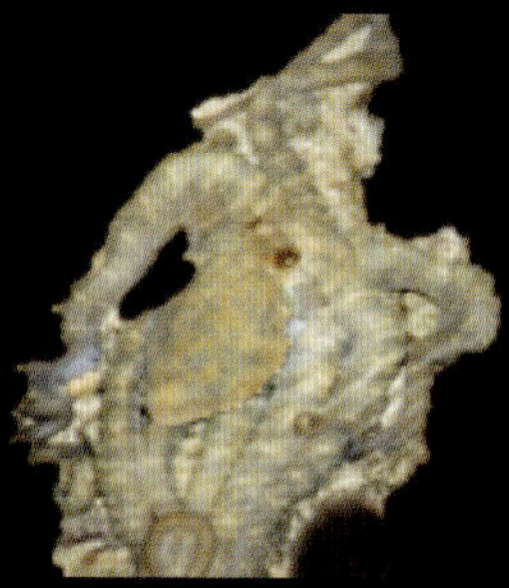

Unidentified Reef Octopus

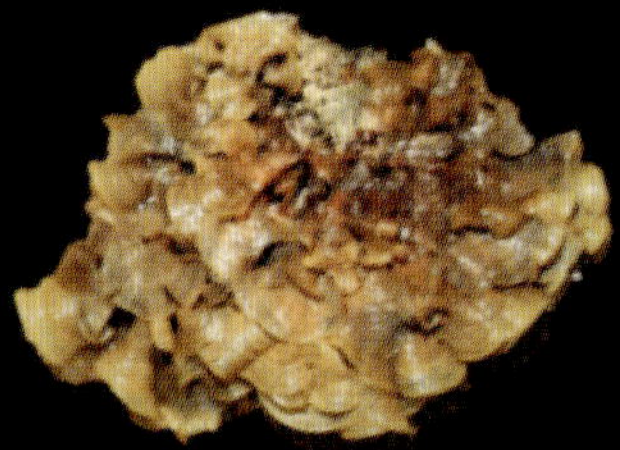

Mermaid's Ear
Padina sp.

Hard Coral
Montipora sp.

Acoporal Coral

Damsel Fish

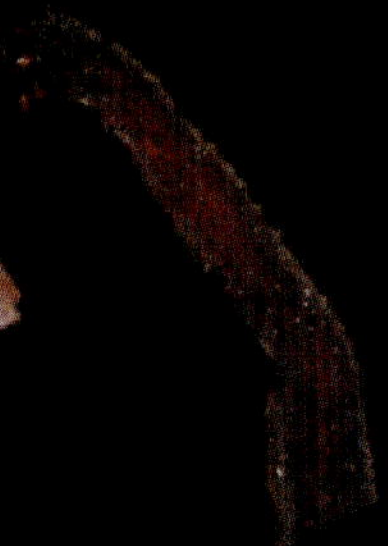

Black Sea-Cucumber
Holothuria leucospitola

Spotted Blenny
Blenniella sp.

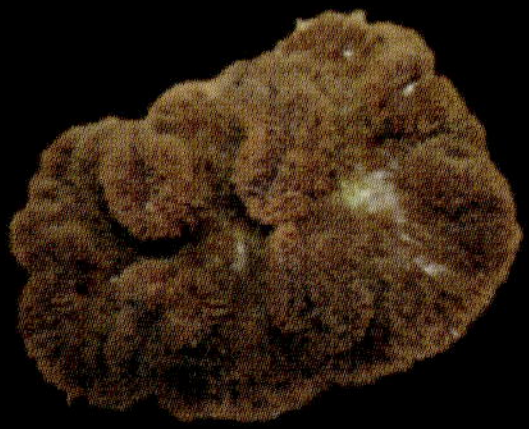

Carpet Anemone
Stichodactyla sp.

Photos and Text:
VINCENT CHOW
Chairman. Malaysian Nature Society Johor branch

CRABS

Crabs are a very large group of arthropods that include lobsters, crayfish, shrimp, krill and barnacles. Many crabs are habitat specific. This makes them excellent bio-indicators of the health and environmental degradation of coral reefs and mangroves.

Batu Batu boasts a healthy repertoire of these crustaceans. Some like the terrestrial land crab live in holes they've dug in the ground. Weighing over a kilogram, they emerge at night to feed on plant debris. The swimming crab is found in tidal pools and is sometimes collected for food by islanders. Another interesting local, the thunder crab is so named because locals believe that if it pinches a prey, it will only let go when there is a clap of thunder.

Ghost crab, Pulau Tengah. (Photo credit: Dave Bakewell)

Thalamita Spinimana

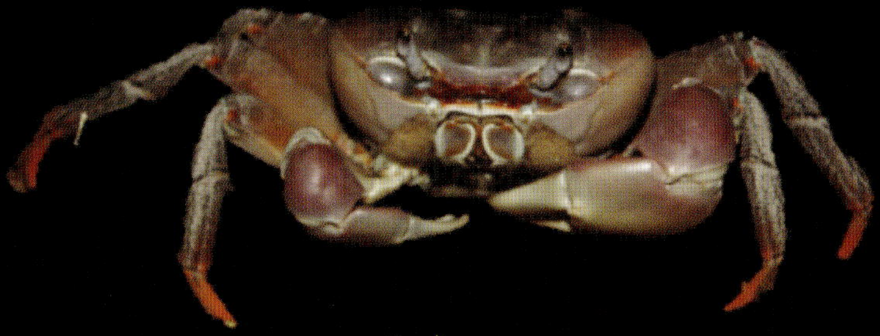

Cardiosoma sp.

Myomenippe Harwickii

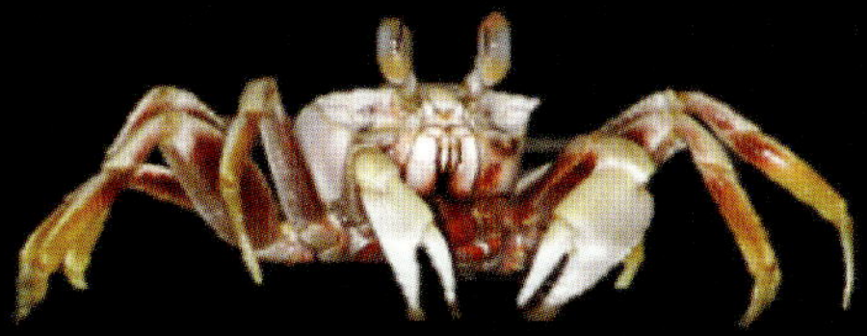

Ocypode Ceratophthalmus

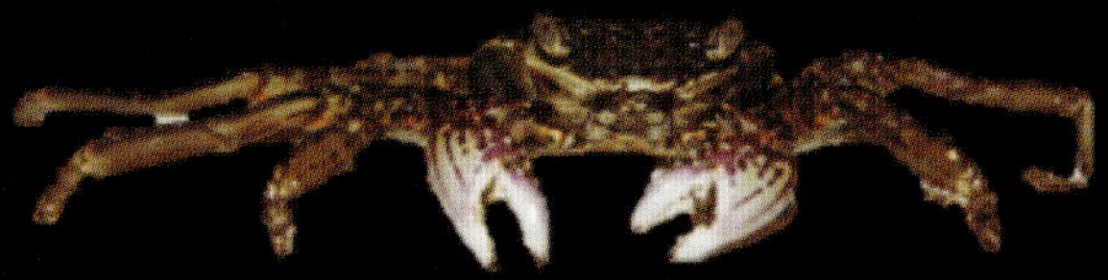

Episesarma sp.

Grapsus Albolineatus

Photos and Text:
VINCENT CHOW
Chairman. Malaysian Nature Society Johor branch

BIRD LIFE

One of Pulau Tengah's great visual joys has long been its fantastic bird life.

This is best appreciated in the early morning, although the patient birdwatcher will be rewarded with glimpses of the island's avifauna throughout the day, regardless of the season.

Resident species such as Common Mynas, Oriental Magpie-Robins, Collared Kingfishers and sunbirds can be encountered around the resort area during daylight hours, and lucky visitors may catch a glimpse of a Large-tailed Nightjar after dusk, or more frequently, hear its monotonous *tonk-tonk* call, which earns it the local name of Blacksmith Bird.

A little exploration along the trails which skirt the coastline can be rewarded with sightings of Olive-winged Bulbuls and Mangrove Whistlers in the scrubby vegetation, while on the quieter beaches, observant eyes may pick out a dark grey Pacific Reef Egret exploring the rock-pools for a snack, or a Common Sandpiper teetering and bobbing its way along the hot sand.

During the migratory seasons, rarer species like Arctic Warblers and Mugimaki Flycatchers from northern Siberia may grace the island for a day or two, before they move on in their epic journeys.

For even the most casual observer, the variety and energy of the sights and sounds of the island's bird life adds an extra dimension to every visit.

As part of its commitment to the conservation of the wildlife of Pulau Tengah, in 2012, Batu Batu commissioned the British ornithologist Simon Buckell to document both the island's bird life and the nearby Mersing estuary, the better to understand the numbers of migratory shorebirds or waders, and other water birds. In all, Buckell recorded 122 different bird species within Mersing, plus another 78 unique species on Pulau Tengah itself.

Buckell's study showed that Pulau Tengah lies on the path of the migratory route for many species, from tiny warblers to mighty honey-buzzards, from ocean seafarers like jaegers and storm-petrels to globe-trotting shorebirds.

He also concluded that the estuary is used as a resting and refueling site for migratory shorebirds returning from their northern breeding grounds on the Himalayan plateaus and the Arctic tundra and mountain ranges before they move onto wetlands further south, as far as Indonesia and Australia.

Left: A Roseate Tern among Black-naped Terns at Tokong Bahara.
(Photo credit: Simon Buckell)

Plain-throated Sunbird

Collared Kingsfisher

Asian Dollarbird

Lesser Frigatebird

Little surprise then that the island attracts a small but appreciative crowd of birdwatchers who arrive throughout the year agog at the possibility of spotting a rare migrant waif or the seldom-glimpsed Nicobar Pigeon.

Spring is an especially fruitful time as the skies above the island are dusted with the migrations of many species under the watchful eyes of patrolling White-bellied Sea-Eagles and Peregrine Falcons. Pacific Swifts dive for insects, Chinese Sparrowhawks drift northwards on swirling thermals, and White-bellied Sea-Eagles and Brahminy Kites move in lazy circles.

Come March and Spotted Doves are cooing sweet nothings in the forest canopy, while Plain-throated Sunbirds cock their heads at the monotonous call of Asian Koels. Resident Asian Dollarbirds and Asian Palm Swifts contemplate nesting just as streams of migratory Barn Swallows and Pacific Swifts power their way northwards over the eastern side of Pulau Tengah.

The annual northward migration continues in earnest in April as a high-pressure system and warm airflow settle over the region. Japanese Sparrowhawks lift up out of the forest and head north, always on the look-out for unwary smaller migrants, such as Arctic Warblers and Eastern Yellow Wagtails. Offshore, other nuances of the same Circle of Life are being played out, as Collared Kingfishers swoop down on scuttling crabs and huge Lesser Frigatebirds patrol the high seas looking for terns to rob of their fishy prey.

Pink-necked Pigeon *(Photo credit: Dave Bakewell)*

Chinese Egret (Vulnerable)

Christmas Island Frigatebird (Critically Endangered)

Nordmann's Greenshank (Endangered) with two Grey Plovers

(All other photo credits: Simon Buckell)

Black-naped Tern speckled eggs

Black-naped Tern chick

White-bellied Sea-Eagle

Malaysian Plover

Common Sandpiper

Great Crested Tern

(All photo credits: Simon Buckell)

Grey-faced Buzzard

By May, the northward migration has slowed to a trickle, as a few straggler Oriental Honey-Buzzards and Grey-faced Buzzards — migratory species of raptors which nest in northern Asia — pass high overhead. Out to sea, around the reefs and rocky outcrops, numbers of Black-naped and Bridled Terns are building up in preparation for the breeding season ahead.

The clear blue skies and hot sunny days of the previous month are now punctuated by thunderstorms in June. None of this perturbs the White-bellied Sea-Eagles soaring over the island, nor the Black-naped Terns who are now guarding speckled eggs — camouflaged against predators by cleverly mimicking the rock colouring — as they begin to hatch.

On the Mersing estuary, Malaysian Plovers make the most of June and July's quiet calm — wispy clouds hovering over the mirror-flat turquoise sea — by nesting and tending to their new hatchlings. The estuary has the sleepy feel of a school vacation: the shorebirds have largely returned to the Himalayan plateaus and Siberian tundra to breed, while the few that remain — a small flock of young Great Knots have decided that migration is not for them this year — have little to do except wait for the advent of the southward migration season and the return of their shorebird comrades.

These begin returning to the area in August. Led by Lesser Sand Plovers and Great Crested Terns, they arrive in ever increasing numbers at the Mersing estuary, hungry from their transcontinental journeying, and in need of the food which the mudflats, exposed at low tide, provide in abundance. Meanwhile, it is breeding time for a thriving colony of Bridled Terns on the nearby rocky islet of Tokong Bahara. This colony was only discovered in 2012 — the first confirmed breeding record of the species in Peninsular Malaysia for several decades — and a reminder of how much is still waiting to be discovered in these waters.

Island Flying-Fox on the banyan tree at Batu Batu beach. *(Photo credit: Simon Buckell)*

September brings with it clear warm sunny days and still more flotillas of shorebirds, which increase in both number and variety. Pacific Golden Plovers and Broad-billed Sandpipers arrive on the increasingly congested shoreline. A change in the temperature of the sea causes an eruption of plankton and huge shoals of fish which, in turn, attract tubenosed Swinhoe's Storm-Petrels — ocean-based seabirds on their winter migration from the seas of Japan to the warmer Indonesian waters.

In October, Bridled Terns patrol the offshore waters, while a rare Chinese Egret, freshly arrived from breeding quarters in Korea or northern China may be discovered high-stepping through the surf in Mersing. The ripening figs of the Banyan Trees lure gourmet Pied Imperial-Pigeons and Pink-necked Pigeons by day, and considerably noisier, squabbling groups of the Island Flying-Fox which are equally fond of the lush, sweet fruit, by night.

By November, the air becomes more charged as the northeast monsoon gathers in the wings. Rising high on strong thermals, the Grey-faced Buzzard heads south. Fully sated, Island Flying-Foxes flee before the choppier seas and increasing rain showers, though small numbers of Arctic Warblers, and Mugimaki and Asian Brown Flycatchers — now well on their southward migratory peregrinations — contemplate staying on in the relatively safe leafy arbor of the woods along the beach.

And finally, the annual avian calendar begins to close. December brings with it the full force of the year-end monsoon, though Pacific Reef Egrets, Kentish Plovers, Marsh Sandpipers and Far Eastern Curlews remain undaunted as they calmly go about their daily business, no doubt already preparing themselves for another busy new year.

Bridled Terns

Brahminy Kite

Great Knots (left), Grey Plover (centre) and Bar-tailed Godwit (right). The godwit is wearing coloured leg bands attached by Chinese scientists near Shanghai which enable researchers to identify key resting sites used by these birds on migration.

Mugimaki Flycatcher

Pacific Reef Egret

(All photo credits: Simon Buckell)

Pied Imperial Pigeon *(Photo credit: Dave Bakewell)*

Lighthouse at Batuan Tikus, diving site and home of the Black-naped Terns. *(Photo credit: Simon Buckell)*

Lesser Sand Plover

Flock of shorebirds at the estuary of Mersing river. *(Both photo credits: Dave Bakewell)*

Sauropus androgynus

Adiantum latifolium

Trimezia martinicensis

Scleria sumatrensis

Podocarpus polystachyus

Artocarpus gomezianus

Calamus burkillianus

Tacca palmata

Lycopodiella cernua

Pronephrium asperum

Ardisia elliptica

Pandanus sandens

BOTANY

One of the reasons why botanists have long been fascinated with islands — Charles Darwin's experience on the Galapagos comes to mind — is that they provide ideal microcosms for studying biodiversity patterns on small island ecosystems, and for understanding the broader evolutionary process. Isolated from the rest of world and, for the most part, undisturbed by the human variable, islands are, for all intents and purposes, time capsules.

The reality, of course, is that the biodiversity of tropical islands have often been adversely affected by human development. In turn, this severely limits the pool of islands that are available for study.

A striking feature of Pulau Tengah is that, despite intensive human intervention in the early 1900s and mid-1970s, at least half of the island has remained relatively intact.

So much so that when Batu Batu opened along the southern shoreline of Pulau Tengah, its owners were sufficiently encouraged to embark on an ambitious programme of conservation and eco-responsible tourism. Part of this commitment included commissioning Dr Bruno Senterre to conduct an exhaustive inventory of the island's flora and vegetation with the goal of establishing conservation priorities and sustainable development.

Senterre's report — completed in December 2013, after an extensive six-week study of Pulau Tengah — was fascinating for its meticulous documentation of the island's botany. For starters, it confirmed what had long been suspected: the island was part of a group off the south-eastern coast of Peninsular Malaysia that is inhabited by species that are either rare or absent from the mainland.

All photos in this chapter by Dr Bruno Senterre

Leaves specimen at Dr Bruno Senterre's lab.

As far back as the 1930s, the English botanist and assistant director of Singapore's Botanic Gardens, EJH Corner had visited the island and recorded 40 plant species. At the time, the northern and eastern parts of the island were still covered with natural forests while the western and southern parts had been planted with coconuts in the early 1900s. In 2012, Quirijn Coolen, a Batu Batu intern, not only recorded 70 species, but also provided the first documentation of the effects of the occupation of the island by 10,000 Vietnamese refugees between 1975 and 1981. During this latter period, most of the secondary forests which had regenerated from the original coconut plantations were again deforested to provide timber for the camps. The original forestation in the north and east of the island has remained mostly untouched.

With the permission of the Johor Parks and Forestry Department, Senterre divided Pulau Tengah into grids using existing paths and trails. Systematically, he combed each grid, collecting samples as he went along. In all, 308 plant specimens were collected and these were placed in a customised dryer for preservation — duplicate samples have been deposited with the Forest Research Institute of Malaysia and the University of Malaya, while a third set is stored on Pulau Tengah itself under the custody of Batu Batu.

Platycerium bifurcatum

Tradescantia spathacea

Clerodendrum myrmecophila

Dicranopteris linearis

Dianella ensifolia

Ptychosperma macarthurii

Metroxylon sagu

Etlingera elatior

Cycas edentata

Dendrobium crumenatum

Tacca leontopetaloides

Schizaea dichotoma

Top: In the field, collecting samples of *Pteridium caudatum subsp. yarrabense*.
Left: Marsh situated just behind the rocky outcrop of North Beach.

Senterre recorded 312 species belonging to 252 genera and 101 families. These include 208 dicotyledons, 70 monocotyledons, four gymnosperms and 30 ferns.

The island's native flora is represented by 216 species comprising 77 trees, 54 terrestrial herbs, 39 shrubs, 21 vines, two lianas and 10 epiphytes. Exotic flora is represented by 96 species comprising 28 trees, 41 terrestrial herbs, 20 shrubs, four vines, two lianas and one epiphyte. A further 57 cultivated exotic species correspond to 22 trees, 13 terrestrial herbs, 18 shrubs, one vine, two lianas and one epiphyte.

Incredibly, Senterre concluded that the natural and semi-natural forests, especially those along Pulau Tengah's east coast, were mostly unaffected by invasive species.

To the untrained eye, Pulau Tengah's terrain may appear as a single homogenous sweep of green vegetation and pristine beach that's punctuated with the island's signature collection of rocks and boulders.

There are, in fact, five distinct eco-zones.

Mangrove Forests — Found on the island's north coast.

Coastal Marshes — Once prevalent on the northern coast, these have generally been converted to dense grassland and water palms as a result of the occupation by the Vietnamese refugees in the mid-1980s.

Coastal Zone — This comprises a narrow belt from the coastline up to around 10 to 30 metres inland. This, in turn, can be broken into three sub-categories: a narrow one- to five metre wide coastal sandy beach dominated by shrubs, creepers and terrestrial herbs; a coastal rocky shore mostly along the north-east part of the island characterised by orchids and dry zone terrestrial species; and coastal forests that grow at the outer edges of the coastal sandy beaches and rocky shores.

Littoral Zone — This extends several kilometres inland and is characterised by beach vegetation and the reduction or absence of typical coastal elements such as the fish poison tree. This is also divided into three sub-categories: mesic littoral forests that cover most of the ridges and slopes of the centre of the island; valley littoral forests teeming with lianas, ferns and palms that stretch along valleys and ravines in a three to ten metre wide band; and dry littoral forests dominated by hardy sea-teak pines, and silverback along the eastern slopes that are exposed to the north-eastern monsoon.

Secondary and anthropic vegetation — this includes pioneer vegetation such as deforested stands which regenerate naturally after human impact or natural tree fall (white-stemmed button vines are a common species); regenerated secondary forests of mostly 10–15m tall sea apple trees on the south-western slopes that were deforested during the years of the Vietnamese refugees; and anthropic areas which have been continuously affected by human intervention and which are characterised by naturalised and invasive exotic species such as downy shieldferns, and Indian goosegrass.

Senterre's study identified several potential threats in the form of exotic species that have, with time, become naturalised to the island. These included the Malaysian sage, devil's ivy, and the climbing hempvine.

That said, the relatively rare absence of exotic flora and vegetation on Pulau Tengah is striking. In an effort to balance education with conservation, Senterre suggested a special protocol be set up that would encourage visitors to the island to explore the botanical trails at the same time as it reduced the spread of exotic species.

In particular, Batu Batu has committed itself to the rehabilitation of areas most visibly influenced by human activity.

Senterre also suggested that new forests be planted to replace the once damaged during the years of the Vietnamese refugee camps. Regular extensive clean-ups are carried out especially along Pulau Tengah's coastline and perimeter of rocks which trap debris, waste and rubbish that are either historical or that have floated in the greater South China Sea.

The task to save and preserve Pulau Tengah's botany and fragile ecological system is an ongoing one. It is a perpetual struggle to balance nature and human activity in a way that provides room for both to prosper. This is, of course, not an easy task, but it is, by the same token, an infinitely rewarding one.

Right: Marsh behind Red Sands beach.

Threats to the biodiversity of Pulau Tengah

Small tropical islands are unique and play an important role in the conservation of biodiversity. Unfortunately, they also tend to be more sensitive to change, especially to the introduction of new exotic species. It is not yet well understood why, but island ecosystems are more adversely affected by exotic species compared to continental areas.

Although just a small fraction of exotic species 'develop' an invasive behaviour, their impact can be devastating especially when they escape the cultivated perimeter and invade wild areas where they can overwhelm the original plants.

As an eco-responsible tourist resort, Batu Batu has engaged in baseline studies to inventory both the native and the introduced flora and fauna of the island. Over and above a simple naturalist curiosity, these studies play an important role in defining priorities in terms of the conservation of the rarest or most threatened native species, as well as for early detection of the presence of invasive species.

Two hundred and sixteen species of plants native to Pulau Tengah have been documented alongside a hundred exotic species. Around 60 of this latter group have been introduced for ornamental reasons, for resort landscaping and gardens, and some as fruit trees. None of these garden varieties has escaped into the wild and no blacklisted invaders have been found among the inventory.

That said, some invasive species have been identified, though these are believed to have been brought to island either accidentally, or they have been carried by birds. These are mostly common weeds that have remained restricted to the open areas occupied by human developments.

Only four of these naturalised species have been able to become invasive to the island's natural ecosystem. The main threat comes from the *Clidemia hirta*, a small shrub originally from South America that has become a global threat to global biodiversity. Already, it has invaded on a massive scale the nearby inhabited Pulau Besar. Thankfully, on Pulau Tengah, only two clumps have been seen in the pristine central valleys, probably brought from seeds by birds visiting from Pulau Besar.

The few other problematic invasive species, such as the *Lantana camara*, have invaded the lands degraded during the time of the Vietnamese refugees; these are mostly on the north-western slopes.

Batu Batu has dedicated itself to programmes to control or eradicate these invasive exotic species, and to promote the conservation of the rarest native species through the restoration of forests on the degraded western and northwestern slopes. It is hoped that such actions will not only contribute to the preservation of Pulau Tengah's biodiversity, but they might also be used to promote public awareness and to spread similar conservation initiatives to other island resorts.

Clidemia hirta is a well known tropical pest. On Pulau Tengah, the few clumps that occur in the wet valleys of the centre, probably brought in by birds, might one day threaten the island's biodiversity.

Lantana camara or bunga tahi ayam tends to invade natural ecosystems. On Pulau Tengah, it is found on the degraded north-western slopes, as a result of the occupation of Vietnamese refugees in the late 1970s.

Clerodendrum myrmecophila is a relatively rare shrub of the forest with branches that provide shelter for ants, and delicate colourful flowers.

Staurogyne setigera, also known as pokok checheh merah, is a small rare herb growing in the shade of the wet valleys in the centre of the island. This is one of the rarest and most beautiful flowers of the forest.

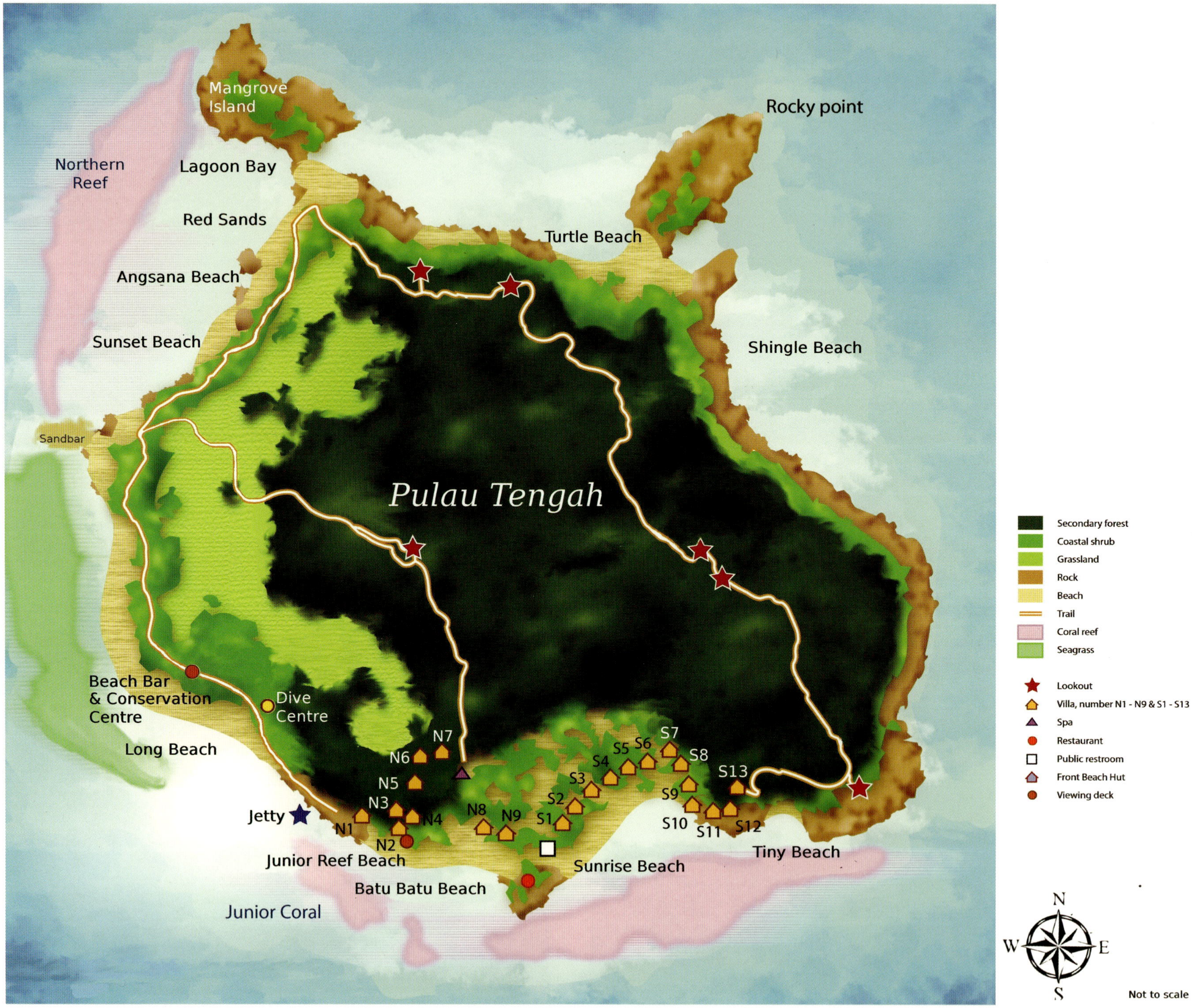

Map of Pulau Tengah showing the main trails and the areas more intensively degraded during the Vietnamese occupation (light green on the left). The dark green area represents partly secondary forest (on the south, west and north-west slopes) and also large areas of primary forest (on the north slopes, in the centre and on the east slopes). *(Photo credits: Batu Batu and Dr Bruno Senterre)*

A WALK ON THE WILD SIDE

Exploring Pulau Tengah's Flora and Vegetation on Foot

by Dr Bruno Senterre

Because they typically have much more limited resources like natural water, small tropical islands tend to have less intense human development compared to larger islands or the mainland. This means flora and fauna species that may have disappeared elsewhere continue to thrive, providing botanists with not just unique pieces of the larger evolutionary puzzle, but also a rare opportunity to conserve rare specimens.

Pulau Tengah is a perfect example. Its human contact has been relatively few and none has been on a large, catastrophic scale. Relatively pristine vegetation still covers about half of the island, mostly concentrated in the central valleys radiating outwards towards the north, east and south-east slopes.

A number of fascinating walking trails present themselves.

All photos in this chapter by Dr Bruno Senterre

View of (left to right) Pulau Hujong, Pulau Harimau and Pulau Mensirip from north Pulau Tengah.

Screw Pines *(Pandanus odoratissimus)* on the south side of the island.

THE WEST TRAIL

Beginning just behind the Batu Batu spa, a well-maintained trail crosses the island and ends at Sunset beach. This trail bears witness to the most obvious signs of deforestation caused during the residence of the Vietnamese refugees from the mid-1970s to the early 1980s.

At the midway point, the view is of the upper northern slopes of the island where deforestation was less intense: here, the fragile soil has been preserved, and regrowth is well in progress. Trees are still relatively small and their very regular sizes are signs of the youth of that secondary forest.

The more pristine forests down the valley are dominated by palms; the vegetation in centre of the island, and the hills of the north-eastern side remain mostly untouched.

Edge of the forest seen from the west-side trail going down the degraded slopes towards Sunset Beach. Here, the impact of the Vietnamese occupation was more intense and affected the soil to such an extent that forest redevelopment is still at an early stage.

Heading further north on the trail, the forest edge starts giving way to the more heavily degraded slopes facing Sunset Beach. There, the original deforestation was the most intense, and the soils were affected in such a way that trees have not yet recolonised. Only large herbs, mixed with clumps of shrubs and isolated trees, dominate. With patience, the forest will eventually redevelop, but it will take more than 50 years before the forest floor is shaded once more, and over a century before the original rare forest species return.

The trail crosses remnants of primary forest. Here, the trees are taller and the understorey is increasingly dense, though the walking remains relatively easy.

The richness of the flora species is reflected by about a hundred different kinds of plants, dazzling in their diversity of leaf shapes and colours. The dominant tree in the canopy is the Krian Batu, a familiar species of the lowland forests near the ocean.

Syzygium grande (family Myrtaceae; local name: Jambu ayer laut, Krian batu). This is the most common tree on the island, mostly found in the interior.

Oncosperma tigillarium (family *Arecaceae*; local name: Nibong). This tall spiny feather palm is common in forests near the sea, and its wood is used as mooring posts.

The large spiny palm tree, the Nibong is also well-known locally for its use in construction and as mooring posts. In the island's central valleys, large clusters are found mixed with other palms, forming an amazing palm forest that is accessible only with a guide. Among these is a species of climbing palm, also remarkably prickly, and belonging to a group of palms used for the construction of rattan furniture.

Calamus burkillianus (family *Arecaceae*; no local name). This climbing palm with strong prickles belongs to a group better known as the rattan and used to make furniture. It is rare and endemic to peninsular Malaysia.

View of the pristine forested slopes of the East coast seen from Rocky Point.

Pandanus odoratissimus (family *Arecaceae*; local names: Pandan duri, Pandan todas). This common tree is found on the well preserved and drier east slopes. Its leaves can be used for making baskets and other home accessories.

THE EAST TRAIL

On the eastern side of the island, the trail continues back to Batu Batu passing by Shark Point Bay. The eastern slopes are the most inhospitable of Pulau Tengah, but they are also among the most interesting areas for the naturalist.

The trail winds through low hills, following the main ridge. On the left hand side, the slopes are often very steep. The vegetation differs from those of the western trail, being drier and more exposed to the north-eastern monsoon. Here, trees are shorter and the undergrowth much denser. While the species richness is less, the area contains a greater number of rarities.

Interesting species along this trail include a larger species of pandan that forms isolated clumps or small-sized populations. Its long prickly leaves can be used to make baskets.

Dischidia major (family *Apocynaceae*; local name: Akar kul, Akar bano). The leaves of this amazing plant have been modified to resemble pitchers which provide accommodation for symbiotic ants which, in turn, protect the plant from parasites.

The drier vegetation tends to be dominated by more prickly species, and also more ants-associated plants, like the *Clerodendrum myrmecophila* whose hollow branches host armies of ants. Another spectacular ants-loving plant is the Akar bano whose pitcher-like leaves have also evolved to host ants.

Fire ants are more common on the eastern slopes, building their nest in the branches of shrubs, and attacking everything approaching. Fortunately, they are found lower down the slope, and not close to the trail.

Away from the trail is found a population of *Cerbera manghas*. As soon as it's gripped or hurt, it releases a mildly toxic white sap.

Before arriving back to Batu Batu, the trail crosses less exposed slopes which feature a forest of large beautiful trees whose light brown bark exfoliate in narrow vertical scales. This tree, the Podo laut, is a tropical cousin of the European conifer which can be trained into bonsai.

Margaritaria indica (family *Euphorbiaceae*; local name: Mala aechil). This coastal tree is remarkable for its reddish bark which it exfoliates in a beautiful pattern.

Terminalia catappa (family *Combretaceae*; local name: Ketapang). This widespread coastal tree, also known as "Indian almond", produces abundant fruits which can be found dry next to the beach.

Scaevola sericea (family *Goodeniaceae*; local name: Ambong-ambong). A beautiful native shrub growing on the sandy beach and flowering throughout the year.

THE SEA SHORES

A narrow fringe of vegetation rings around Pulau Tengah before extending almost 10m inland. Here, the vegetation has adapted to the stress of sea spray and difficult soil conditions that are sandy and salty or brackish.

Many species here are not found further inland. Among them, *Margaritaria indica* is one of the rarest species on the island. This remarkable, large tree is easily recognized by its reddish bark. It can be seen just before arriving at Long Beach, and several other areas along the north coast. Unlike most plants, trees of this species have separate sexes: some trees have only male flowers, others have only female. This might partly explain the rarity of the plant as colonisation of new sites typically require both a male and a female tree to be effective.

On Long Beach, and further north on Sunset Beach, it's tempting to role-play Robinson Crusoe by eating coconuts from the abandoned plantations, or test one's dexterity opening the dried fruits of the Terminalia catappa to get to its delicious Indian almonds which taste a little of hazelnuts.

The beautiful flowers of the coastal shrub *Scaevola sericea* are used abundantly in Batu Batu's landscaping; the plant is prized for its capacity to form dense vegetation walls near the sea.

Casuarina equisetifolia (family Casuarinaceae; local name: Ru, Aru, Eru). Featuring prickly fruit, this tree is typical of the coastal forest, especially on sandy soils. It protects the island from the wind and sea sprays.

The dried prickly fruits of *Casuarina equisetifolia* can be worked in many ways to create decorative objects and necklaces.

Hibiscus tiliaceus (family Malvaceae; local name: Baru, Ambaru, Baru laut, Dedap laut). This is a very typical coastal tree whose dried fruit is also an interesting decorative object.

The dried prickly fruits of the filao tree makes life difficult for the barefoot explorer though locals love to use it to make beautiful necklaces and other woodcrafts. Some of deadwood found on the beach, washed and polished by the sea, are very decorative.

As the path meanders towards the north coast, the coastal vegetation fringe becomes more natural. Coastal hibiscus trees are common and shed their large colourful flowers on the beach.

Some mangrove forests, an intermediary between terrestrial and marine ecosystems, can be found on Small Island and Mangrove Island. These include just a few species that have evolved specialised root structures to cope with the fluctuating tides and the salinity of substrate in the mangrove. Some have large arched roots coming out of the trunk while others have roots growing upward rather than downwards.

Rhizophora mucronata (family *Rhizophoraceae*); local name: Belukap, Bakau kurap). This is one of the few mangrove tree species. Its large aerial arched roots provide stability to resist the waves, and also to retain sediments and improve access to air for respiration of the roots.

View of the steep eastern rocky slopes where specialised orchids grow on the rocks.

Top: *Cycas edentata* (family *Cycadaceae*; local name: Paku laut). This coastal tree fern is a living fossil that belongs to a very old group of plants. It is more related to pines and *Araucaria*.

Right: *Cycas edentata* seen in its habitat of rocky shores.

View from Rocky Point towards the pristine north-eastern side of the island, with the steep rocky slopes (on the left) and the mangrove forest (in the front).

On the eastern coast, where the slopes can be rockier and steeper, the botanical marvels are even more stunning. In this relatively inaccessible area, one of the most remarkable species is the *Cycas edentata*. This plant is much valued around the world for its ornamental properties. Extremely slow growing, it appears to be an intermediate between ferns and palms; the leaves unfold as a cross but bear true seeds rather than spores. *Cycas* are a kind of living fossil, and they were already present at the time of the dinosaurs.

The east coast is also the only place on the island where several populations of rare orchids are found. Some are large terrestrial herbs like *Cymbidium finlaysonianum*; others like *Arachnis hookeriana* typically grow on rock faces.

Sonneratia griffithii (family *Lythraceae*; local names: Bedada, Pedada, Gadabu, Perepat). A mangrove tree species with upwardly growing roots to allow respiration.

Arachnis hookeriana (family *Orchidaceae*). Found on the rocky shore of the exposed east coast of the island, this relatively rare orchid grows on rock boulders. It requires conservation action.

View from the north-west of the island, right to left, Pulau Rawa, Pualu Gual, Pulau Mensirip and Pulau Harimau. *(Photo credit: Batu Batu)*

View of Pulau Rawa from the south-east of the island. *(Photo credit: Batu Batu)*

THE VIETNAMESE BOAT PEOPLE

In 1975, after several decades of a fretful and complicated civil war, the North Vietnamese communist forces finally overcame Saigon. In the ensuing turmoil and panic, entire generations fled their homes in a mass exodus. Those unable to leave by land or air took to the sea in ragged armadas of smuggler boats and chartered leaky fishing boats.

In the main, this Vietnamese diaspora of refugees comprised ethnic Chinese businessmen and intellectuals and their families.

The boats carrying these desperate refugees (which later included waves from Laos and Cambodia) fanned out across the Gulf of Thailand and the South China Sea towards Malaysia, Indonesia, Hong Kong, Thailand and Australia. The journeys from Vietnam were perilous and a good many refugees perished from a combination of dehydration, disease, storms, hunger and opportunistic sea pirates.

Those lucky enough to make landfall faced months, sometimes years, in refugee processing camps run by the United Nations and various Australasian governments.

Incredibly, a great many landed in Malaysia, though mainly on two east coast islands in the South China Sea where the transit camps were managed by the United Nations High Commissioner for Refugees.

Between 1975 and 1991, Pulau Bidong, Trengganu took in around 250,000 refugees. More importantly for our purposes, Pulau Tengah was, from the mid-1970s to early 1980s, a United Nations' Refugee Council Transit Camp for around 120,000 refugees.

Here, on the western coast of this sea-wet rock framed by white-sanded beaches and palm trees — an unexpected paradise after their ordeal at sea — this traumatised diaspora waited for the call to come from their new homes where they could start afresh with new lives.

Left: Vietnamese huts at Sunrise Beach, circa 1978.

Many lives were lost during the perilous sea voyages in leaking boats.

Exhausted after a long journey at sea, a dangerously overcrowded refugee boat arrives at Pulau Tengah, circa 1978.

Eventually, the United States accepted from the greater Vietnamese refugee community a total of 823,000 refugees, United Kingdom 19,000, while France, Australia and Canada took 96,000 each.

The presence of the refugees had a large, long-lasting impact on Pulau Tengah's ecology, the effects of which are still visible today. Vast tracts of forest were denuded for firewood and building purposes. The Malaysian government, which had taken over the island temporarily for the purposes of the camp, paid the island's owner compensation of five cents for every coconut tree that was felled.

There were three camps on Pulau Tengah. Long Beach was then called Trung Dao — easily identified by the big abandoned steel ship called Yet Kieu on the sand; Angsana Beach and Red Sand Beach, Bac Dao; and Sunrise Beach, Nam Tao.

Most of the daily activities were focused on Long Beach where the main buildings were located including a main hall, kitchen, concrete wells, the Malaysian police department and toilets.

Newly arrived refugees were interviewed in the main hall which was also the medical centre. Today, the concrete remains are still visible next to Batu Batu's dive centre. Each refugee was photographed and given an alphanumeric identity that consisted of the letters of the place they'd come from in Vietnam and the number of the boat that had brought them to the island.

During the day, most of the refugees wandered on the beach, congregated in front of their homes, or dipped in the sea. It was rare to see anyone fully immersed in the water as few knew how to swim.

(Photo credit: Kevin Gryce and Anne Jeanneret)

UN personnel would distribute food on Long Beach — condensed milk, tins of sardine, vegetables, eggs, tea, cooking oil and rice. There was a small provision shop where those who had money could buy fruits and soap.

Water was drawn from dozens of dug wells scattered all over the island and from Long Beach's three concrete culvert wells, and boiled before use. There were two jetties on Long Beach — one was used to moor the boats, and the other for toilets and showers.

Every morning, UN personnel came to check on the camp. Nobody else was allowed to visit Pulau Tengah, not even the owner.

Years later, long after they had been repatriated, a number of the refugee families returned to Pulau Tengah. Invariably, their return was a mix of bitter-sweet nostalgia and astonishment at the physical transformation of the island.

What struck those returning was just how few physical traces were left of the original camp buildings. For the most part, makeshift housing and camps that once housed thousands had been dismantled. Other than the remnants on the northern beaches of brick bases, and the three concrete wells at Long Beach (the latter are still in excellent condition and in use today by Batu Batu), the jungle had reclaimed the rest, camouflaging the past with exuberant emerald-green vegetation.

UNHCR calvert wells at Long Beach.

Left: Excavation of an old refugee boat at Long Beach in 2011.

Game of soccer on Long Beach, then known as Trung Dao, circa 1978.

Group on Red Sand Beach, circa 1978.

A refugee boat docks at Long Beach, circa 1978.

Vietnamese refugees arriving at Pulau Tengah.

A funeral procession.

Refugee family resting in their hill hut.

Remains of a refugee camp latrine.

Lang, a refugee (in yellow jumpsuit) as a child with her mother at Long Beach.

Lang as a child with her mother at Long Beach jetty.

The 3m-long boat that Lang and her family spent three days and three nights on when they escaped Vietnam.

Vietnamese family on a fishing bost on Long Beach.

(All photo credits on this page: Lang)

Lang with members of her family on Long Beach.

Typical refugee living quarters (above and below).

Lang and a relative.

Above and below: Graffiti carved onto boulders by the refugees.

UNHCR and Vietnam embassy officials.

UNHCR and Vietnam embassy officials.

UNHCR and Vietnam embassy officials.

Ticson Mach's father, Mach Quang (left) and friend waiting for a UN delivery of food. *(Photo credit: Ticson Mach)*

UNHCR and Vietnam embassy officials.

Special refugee flight tickets. *(Photo credit: Lang)*

Top: From left to right, Trinh's father, uncle, sister and mother at the grave of baby Joseph in 1978 and 2014 (bottom).
(Photo credits: Trinh Nguyen)

I was born in the US after my family had escaped from Vietnam.

Like so many of their generation, my family's journey to freedom was filled with many challenges and heartache.

The elaborate plan to leave a now communist country took more than a year to devise. Precaution was needed as my father's family owned a fishing boat, so they were constantly watched by the new authorities as all families with boats were suspected of trying to escape.

One day in 1978, a small boat was hired to take my mother and my brother and sister to meet my father, my grandfather and uncles who were offshore on the pretext of fishing. They waited at the shore. The boat never came. News travels fast in small villages and word reached the authorities quickly. With the police onto them, the group started walking out farther into the water. Panic set in as the water started rising. My mother had to carry the children on her shoulders.

They heard shooting. A fist was shoved into my sister's mouth to muffle her cries. Meanwhile, my father waited anxiously in his boat. Eventually, worried about how late it was getting, he began heading back towards the shore in complete darkness. By some miracle, he stumbled on my mother. Everyone was pulled on deck.

My parents said the police only allowed fishing boats to have three days' supply of food and fuel, so, as they were leaving, they stopped by their friends' boats to ask for additional supplies. Some friends decided at that point to jump on and leave as well.

The boat eventually set off carrying 24 people onboard: My reunited parents, grandparents, brother, sister, four uncles and family friends.

With no map or navigation system, they were at sea for four days and four nights before land was spotted. When they first tried to dock, a boat that they suspect belonged to the Malaysian navy pulled them back out. They believe aid wasn't provided because their boat was still in working condition. A second landing attempt at a different location was successful. Learning from the previous attempt, they destroyed the engine.

Once onshore, everyone was given food and clothing. About two weeks later, they were moved to the refugee camp on Pulau Tengah.

About a month after settling on the island, my three year-old brother Joseph passed away from an undiagnosed illness. He was the first person from the camp to be buried on the island. Everyone from fellow refugees to aid workers pitched in to help my family in. A stone was constructed to mark the burial site. This proved to be invaluable as it helped us locate his remains 36 years later when my parents returned to the island.

After my brother's death, my family's relocation process was expedited. My parents established a new life in the US. California became their new and permanent home where they had four more children (including myself) and started a successful food truck business.

Trinh T. Nguyen

The funeral procession for baby Joseph.
(Photo credit: Trinh Nguyen)

Ticson Mach and his family on a beach at the Pulau Tengah camp

Ticson Mach and his family today *(Both photo credits: Ticson Mach)*

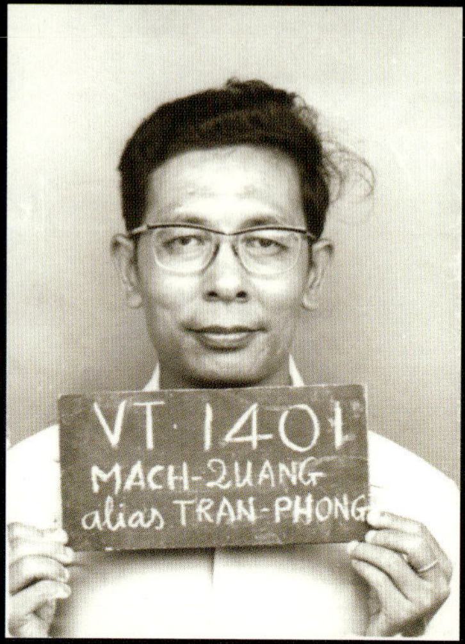
VT 1401
MACH-QUANG
alias TRAN-PHONG

VT 1401
LUONG-NGOC-CHANH

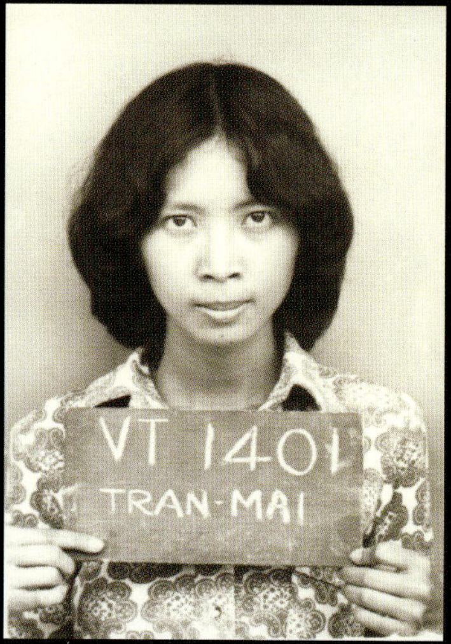
VT 1401
TRAN-MAI

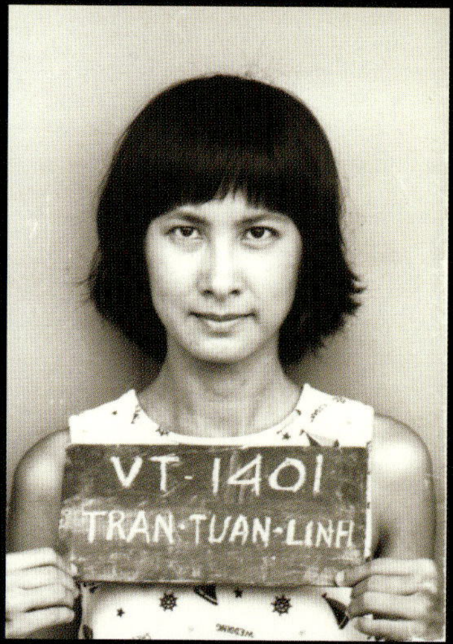
VT-1401
TRAN-TUAN-LINH

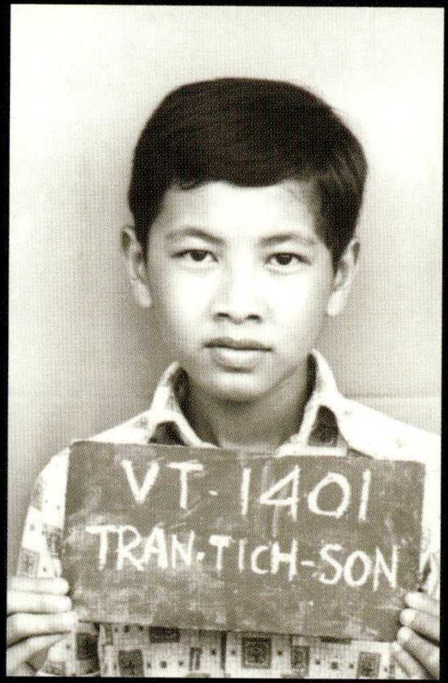
VT-1401
TRAN-TICH-SON

Our family of five fled aboard a fishing boat out of Vung Tau, a seaside town about five kilometres south of Saigon. For four nights and three days, we encountered shark attacks, pirates attempting to board our boat. We had no food and very little to drink.

Miraculously, we found our way to Pulau Tengah. We were so afraid that the Malaysian government would not allow us to land, so, as the boat got closer to the pier, we jumped off and swam onto the beach.

That was September of 1978.

When we arrived, to our surprise, we found my aunt and her family were also on the same island. At first, we squeezed into their house, until we found someone who wanted to sell theirs, because they were leaving to be repatriated to another country, and we bought it from them. At that time, there were an influx of boat people, and so houses were in demand. I believed we bought our house for five pieces of gold.

On a typical morning, after breakfast, we would come down to the main beach to wait for food to be distributed by the United Nations. The food was delivered by boat in baskets full of rice, bell fishes, and vegetables. We were grouped by our boat numbers. The food distributed to each group, and we then divided it amongst ourselves. There was never any chicken, pork, or beef.

After lunch, we would wait in line for water, or go into the jungle to collect firewood and chop it. If it was low tide, we would go to the beach to collect sea cucumber, and clams to supplement our meals.

At night, there wasn't much to do since there was no electricity. We just sat around and talked, maybe played some game. Many of us kids would climb down to the rocky beach, and start yelling at the ocean to release our frustration and boredom.

Although life was difficult, everyone on the island was happy, because we all knew that freedom was so near.

We waited on the island for ten months. After we registered ourselves with the UN, we would wait each day to be called in for an interview. There were many phases. They wanted to know who we were, they checked our background and our health, determined where we had relatives (in the US, France and so on), worked with our sponsor country. Finally, we got approval from the US to take us as refugees through the sponsorship of the local church in Seattle, Washington. We got our date to leave the island.

We were sent to Mersing for a thorough health check, and wait for the departure date. We stayed in Mersing for about a month. Then we were bused to Kuala Lumpur, where we were held for another month before we finally left for the US.

Ticson Mach

(All photo credits: Ticson Mach)

Back row (left to right), Hao Lam (Howie); his mother, Chen Quach; and his father, Thanh Lam. Front row (left to right): Hao's sister, Nga Lam (Nikole), and his brother, Binh Lam (Ben). Chinese New Year, 1979.

From left: Nikole Lam, friend and Ben Lam, Chinese New Year, 1979.

Hao Lam and friends, 1979. *(All photo credits: Howie Lam)*

"My parents were among those who decided to risk everything and flee Vietnam. I was 15. After months of planning, our journey began in November 1978. We boarded a 15-metre long fishing boat with 153 people on board and headed to open seas. Our boat and two other similar boats departed together, but after a few hours we lost sight of one another. Our journey took four days and three nights.

My grandmother died on the voyage and was buried at sea.

One night, we were hit by a tropical storm with high winds and heavy rain. Big waves hit the deck. We were all soaked in water. I felt every bit of the cold howling wind. It was pitch black outside everywhere I looked. All I could hear were big waves crashing against our boat and flooding the deck. People huddled together in fear of what might happen next. I remember hearing people around me praying. Somehow, we made it through.

We saw a local fishing boat and waved for help. The fishermen pointed us towards a chain of islands. One of those islands was Pulau Tengah. While the local authority and representatives from our boat were negotiating our arrival, we were left floating on our boat overlooking what is, today, Long Beach. Hundreds of little shacks were lined up and down the hillside

My family bought a makeshift shelter from refugees who were being resettled. Shelters were no more than tree branches cut down from the trees on the island itself and some boards held together with some blue nylon tarp. The few barracks built by the UN had been filled up quickly, so people built their own shelters.

Our shack which we shared with my extended family of 14, and two other families was just up the hill in the central island section near the four fresh water wells that still remain today.

Life on the island was rather idle. Almost every day, I went down to the beach to cool off. One of my most difficult, yet memorable, moments was to hike up the hill to chop down trees, cut them into logs, and carry them home to make fire for cooking. Every few days, food supply boats arrived from Mersing. Our boat number, BX388, would eventually be called and we would then swim out to retrieve our supplies as they were thrown overboard.

We stayed on Pulau Tengah for four months, a relatively short time but it was a memorable, life-changing part of my life that I will never forget. My family and my aunts were accepted for permanent settlement in America.

My parents have now retired. My younger sister and brother, and I have all finished college and have good careers. All of us are also raising our own families now.

I returned to the island in 2007. What an emotional time it was for me. I was in tears when I arrived. So much had changed. There were so many things I was hoping to find, but they are no longer there. No more huts. The rusted remnants of the hull of the steel boat 'Yet Kieu' that was beached on Long Beach is also no longer there. But to my surprise, the three wells above Long Beach have somehow defied time and look just like they did 30 years ago. These wells were the life-line of the refugees. I hope they will be preserved as a piece of history."

Hao Lam (Howie)

The contestants vote at an elimination round.

EXPEDITION ROBINSON

In 1997, Strix, a Swedish television production house, was given the green-light by a consortium of European television studios to launch a new reality show that would be broadcast in Sweden, Denmark, Holland, Belgium, Norway, Finland and Greece.

The show was called Expedition Robinson, an homage to Robert Louis Stevenson's classic Robinson Crusoe about a man shipwrecked on a deserted island, and to JD Wyss's similarly themed The Swiss Family Robinson.

And like the novels, the show's premise was simple enough. Take a group of 16 ordinary men and women aged between 18 and 65, and maroon them on a deserted island with only the most rudimentary tools and their wits, their only goal being to survive both the island and one another. To do so, alliances were forged. Survival tactics — both literal and metaphorical — became critical because at the end of each episode, one member was voted off the island, with the last contestant standing winning the event and a cash prize.

Besides picking contestants with the right combination of telegenic personalities, psychological steel, physical fitness and survival smarts, it was important to set the show on the right island.

Strix's hunt took them all over the world and ultimately to Pulau Tengah. The logo of the show even had, as its background, a map of the island.

From the start, the show — a beguiling mix of nature documentary, anthropological study of group behaviour under conditions of isolation and stress, and old-fashioned game show — was a bona fide hit at home and abroad. (A little known fact is that the show's format had originally been developed in 1994 by Charlie Parsons for Planet 24, a UK TV production company called Planet 24, but Expedition Robinson was the first out of gate making its debut in 1997.)

(All photos in this chapter by Batu Batu)

Contestants taking a break during an obstacle course at Sunrise Beach.

Obstacle course set up over the water at Lagoon Bay.

Expedition Robinson was the original Hunger Games. A combination of clever editing and casting made for riveting viewing. Each week, literally millions tuned in for their dose of vicarious thrills, cheering as a real life Robinson Crusoe drama unfolded right in their living rooms. The television rights were picked up in 13 countries including Sweden, Norway, Denmark, Greece, Germany, Belgium, the Netherlands, the Baltic countries and Russia. It spawned countless versions around the world, including the much-loved American version, Survivor which debuted in 2008.

At the height of its fame, Expedition Robinson averaged viewership numbers of around 50 million a week. The final episode of the fourth season in 2000 was viewed by one in two Swedes. In raw numbers, that translated into a viewership of just over four million, a staggering statistic by any count.

Not all the seasons of Expedition Robinson were filmed on Pulau Tengah, though one season of Koh Lanta, the French equivalent of the show, was filmed on Long Beach in the summer of 2012.

What's worth noting though is that the very first impression of the island — rocky, white-sanded beaches fringing a lush emerald interior — left such an indelible imprint, it informed the choice

Top: Camera crew in action.
Left: A team challenge on Sunrise Beach.

of locations for subsequent seasons and franchises. In other words, Pulau Tengah became, in the minds of the global audience, a short-hand reference point for the ideal of a deserted island paradise — even when the version they were watching had been filmed on another island altogether.

From the start, the producers of Expedition Robinson were careful that their presence on Pulau Tengah was as delicate as possible. After all, if part of the intention was for audiences to consider the relationship with nature and the environment, it was important that neither film and production crew nor contestants did any irreparable damage to the fragile ecosystem. Indeed, the prime directive was that at the end of the fifty-plus days shoot, no trace would be left of the production.

Batu Batu *(Photo credit: Erik Fearn)*

Batu Batu Beach (Photo credit: Erik Fearn)

BATU BATU

Pulau Tengah, Johor

Opened in April 2012, Batu Batu is a simple, family-friendly resort put in place to allow its visitors to enjoy the island and its natural wonders. It was imperative that the resort enhance the natural beauty of the island rather than detract from it and to this end, Batu Batu's architectural brief was deliberately kept low-key.

Each of the 20 one-bedroom and 2 two-bedroom villas is gently set into the island's wild and natural environment, so as to capture the best views while leaving as much of the natural setting undisturbed as possible.

Built by local carpenters using traditional methods in the classic style of a kampung, a traditional Malay village, every nail, every piece of harvested plantation timber, bag of concrete and sand for the resort was barged onto the island from the nearest port, Endau.

Lifted off the ground by stilts, the villas are well ventilated while preserving the natural contours of the ground. The airy interiors of the villas are clad in natural seasoned timber that frames canopy beds, high ceilings and generously proportioned verandahs that provide perfect views of the sunset.

Meanwhile, the open-sided pavilion restaurant and bar feature panoramic 270 degree views of the ocean.

The villas feature high ceilings with recycled teak furniture. *(All photo credits: Erik Fearn)*

Veranda leading directly down to the beach.

The contemporary design continues in the expansive bathrooms.

Top: Generous deep roof overhangs provide shelter from the sun and the rain.
Left: Restaurant and bar flanked by Batu Batu Beach on the left and Sunrise Beach on the right.

Designed with a simple palette by Y Architects of Johor Baru, the details were kept deliberately spare and practical for ease of construction given the lack of motorized machinery used during construction. The bar tops and buffet tables were constructed from slices of a fallen cengal (or Malaysian ironwood tree) and the Beach Bar stools from chunks of casuarina trunk harvested from the island's fallen trees.

Passing almost unnoticed by most guests, a number of green initiatives are woven into the daily routine. Sewage and grey water are treated by a specially designed water treatment system rather than being released directly into the sea. LED lighting, used through most of the resort, significantly reduces energy use.

(Both photo credits: Erik Fearn)

All villas incorporate solar water heaters. Guests recline on FSC-certified recycled teak furniture, while the bathrooms are stocked with paraben-free toiletries.

As the only development on Pulau Tengah, Batu Batu takes seriously its custodial role of preserving and enhancing the island's natural habitat and surroundings.

The mantra — to tread lightly, source sustainably, behave responsibly and support local business wherever possible — is a philosophy that guides the day-to-day decisions of the resort and the business it generates. This includes sourcing its supplies, services, inventory and staff to the fullest extent possible from the local mainland town of Mersing.

Where possible recycled timber was used in the construction of Batu Batu. A cengal log, *top left*, before being cut into buffet tables, *top right*.

Batu Batu uses a number of energy saving measures including, above, solar powered heaters.

Construction of Batu Batu.

Jungle villas. (Photo credit: Y P Yap)

Hashim, a local carpenter, on site during construction of Batu Batu.

Construction of Batu Batu.

Top: Beach bar at Long Beach.
Right: Villas flanked by Junior Reef Beach (left) and Batu Batu Beach (right). *(Photo credit: Erik Fearn)*

Fully recognizing that sustainable eco-practices are not always easy to adopt, let alone on a tiny, uninhabited island, Batu Batu makes concerted efforts towards becoming more environmentally conscious and responsible as the resort matures. Batu Batu believes strongly the old idea that tourism destroys nature is out-dated and the aim could and should be a harmonious and mutually beneficial relationship between the two.

As the previous chapters show, in the years since Batu Batu opened, the resort has embarked on a series of ecological and conservation projects starting with the thorough cataloguing of first the island and its surrounds, birds and later the island's flora. These projects are carried out on the resort's own initiative with the participation of field experts.

Batu Batu has also collaborated with the Department of Marine Parks to train its dive instructors as Reef Check Eco divers. In exchange, Batu Batu divers carry out Reef Check surveys and collect data from around the island to better understand the state of the coral reefs and marine life in Johor's fragile marine ecosystem.

The beaches — especially those near Batu-Batu — attract Green and Hawksbill turtles that land to lay their eggs in nests from March to October. Batu Batu works proactively with Marine Parks of Malaysia on a Turtle Nest Protection and Hatchery Programme with the goal of nurturing and protecting this small but thriving community of turtles from the vagaries of nature and, more endemically, opportunistic poachers.

Swimming pool by Batu Batu Beach.

The open-air bar at Batu Batu Beach. *(Both photo credits: Amos Wong)*

(Photo credit: Amos Wong)

(Photo credit: Erik Fearn)

TWC
TURTLE WATCH CAMP

Top: Beneath the crystal clear waters surrounding Pulau Tengah is a secret world of marine life and sea-grass meadows.
Left: Turtle Watch Camp @ Batu Batu. *(Photo credit: Erik Fearn)*

Over 2012 and 2013, Batu Batu recruited Simon Buckell, an ornithologist to document the bird life and other fauna of the island. In July 2013, two dive instructors Kevin Gryce and Anne Jeanneret joined the Nature @ Batu Batu team and began to inventory and photograph the island's marine life. They also spearheaded the Turtle Conservation programme for two seasons and saw to the set-up of Batu Batu's turtle hatchery.

Later that same year, Dr. Bruno Senterre, a Belgian tropical botanist, was commissioned to conduct a comprehensive check-list and study of the island's flora which took over one month of fieldwork to complete. Carried out in conjunction with the Forest Research Institute of Malaysia (FRIM), the study yielded a comprehensive and valuable database of not just the island's flora including endangered and rare species, but also created herbariums deposited both at FRIM's depository as well as at Batu Batu.

In 2015, Batu Batu added two new dive instructors with a strong passion for marine conservation, Adam Beard and Carmen Toanchina, to the Nature @ Batu Batu team. Batu Batu has further strengthened its relationship between the Marine Park and the Fisheries Department and set-up the volunteer-driven Turtle Watch Camp @ Batu Batu with Adam and Carmen at its helm in 2015. Since then, TWC has brought

The rugged rocks and boulders of Pulau Tengah after which Batu Batu is named.

thorugh various marine biologists and volunteers who have made turtle conservation on the island possible.

Of course, none of these projects can be carried simply on the backs of good intentions and goodwill. The financial reserves come from a number of sources including a primary revenue stream from the resort which is channelled towards conservation and environmental initiatives on and around Pulau Tengah under the Turtle Watch Camp @ Batu Batu banner.

The quest to rescue and to preserve Pulau Tengah and its surrounds is an ongoing challenge. The island's fragile ecosystem requires constant vigilance. Having started from nothing, much has already been accomplished through the dedication and commitment of uncounted staff, field experts and even guests of Batu Batu. Each has, in their own way, contributed immeasurably to the documentation, analysis, recording and oral memory of the island's birds, plants and marine life. This book is a tribute to them all.

Beneath the crystal clear waters surrounding Pulau Tengah is a secret world of marine life and sea-grass meadows.

Sunrise Beach *(Photo credit: Ahmad Massakino)*

Sunset at the pool. *(Photo credit: Erik Fearn)*

APPENDICES

SEMENANJUNG MALAYSIA
JOHOR
Skala (Scale) 1: 200 000
PAHANG
PULAU PEMANGGIL
PULAU AUR
PULAU BABI
PULAU TINGGI
PULAU SIBU
(MUKIM TIOMAN)
(MUKIM PULAU BABI)
LAUT CHINA SELATAN
(SOUTH CHINA SEA)
TERIANG
PENYABONG
PDG. ENDAU
SEMBRONG
TENGLU
BEKOK
MERSING
JEMALUANG
KAHANG
PALOH
NIYOR
LENGGOR
TENGGAROH
KLUANG
RENGAM
KAMBAU
ULU SUNGAI SEDILI BESAR
ULU SG. JOHOR
SEDILI BESAR
LAYANG LAYANG
MACHAP
KOTA TINGGI
ULU BENUT
SEDENAK
KULAI JAYA
KULAI
SEDILI KECHIL
KELLANG
BENUT
BUKIT BATU
SG. PINGGAN
AYER BALOI
PONTIAN
JOHOR LAMA
SENAI
JOHOR BAHRU
TEBRAU
SG. TIRAM
TG. SURAT
PONTIAN
PLENTONG
PANTAI TIMUR
PENGERANG
TG. KUPANG
SG. KARANG
SERKAT
SINGAPORE
SELAT SINGAPORE
(STRAITS OF SINGAPORE)

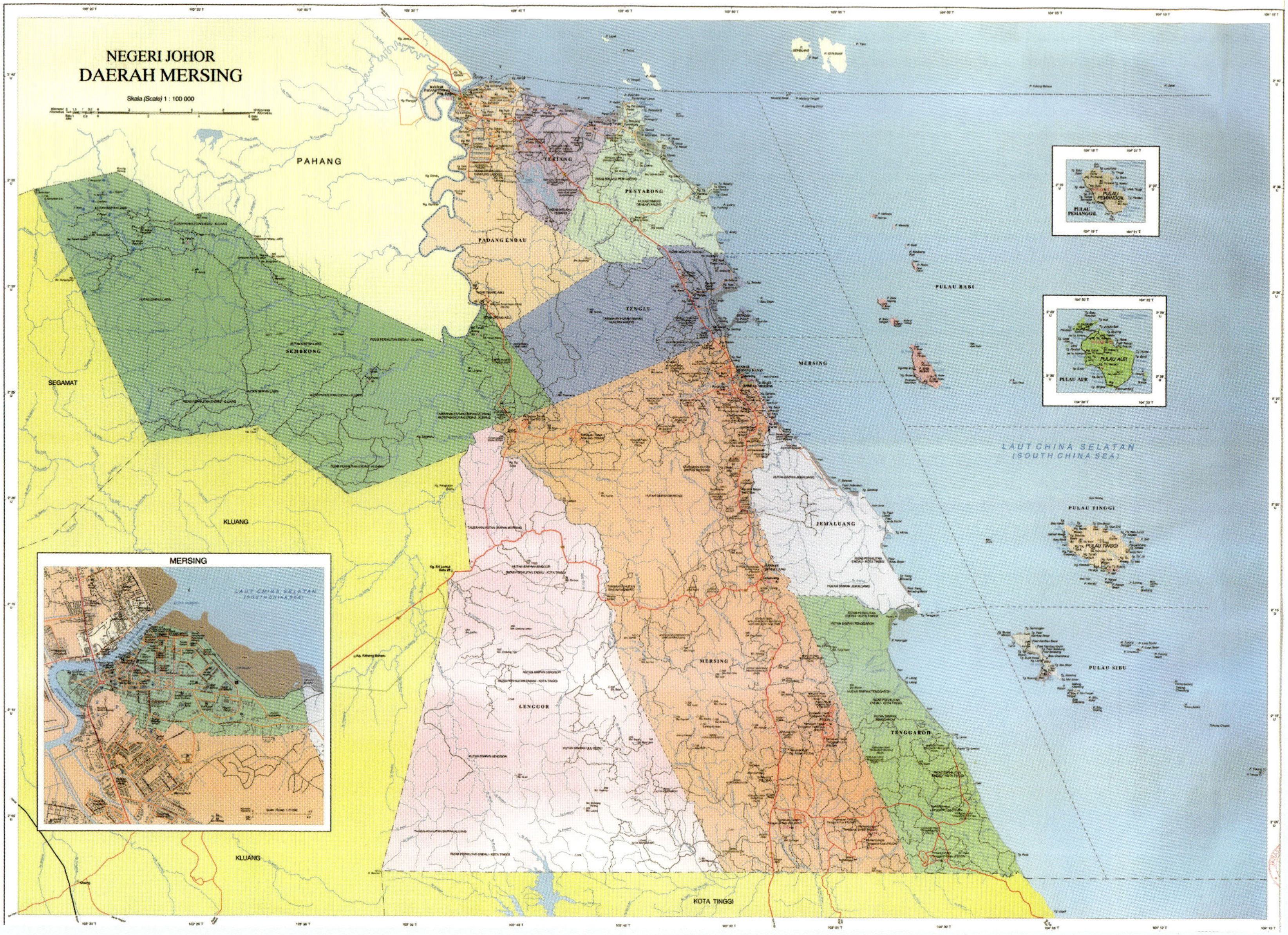

(All maps courtesy of Jabatan Ukur & Permataan Malaysia)

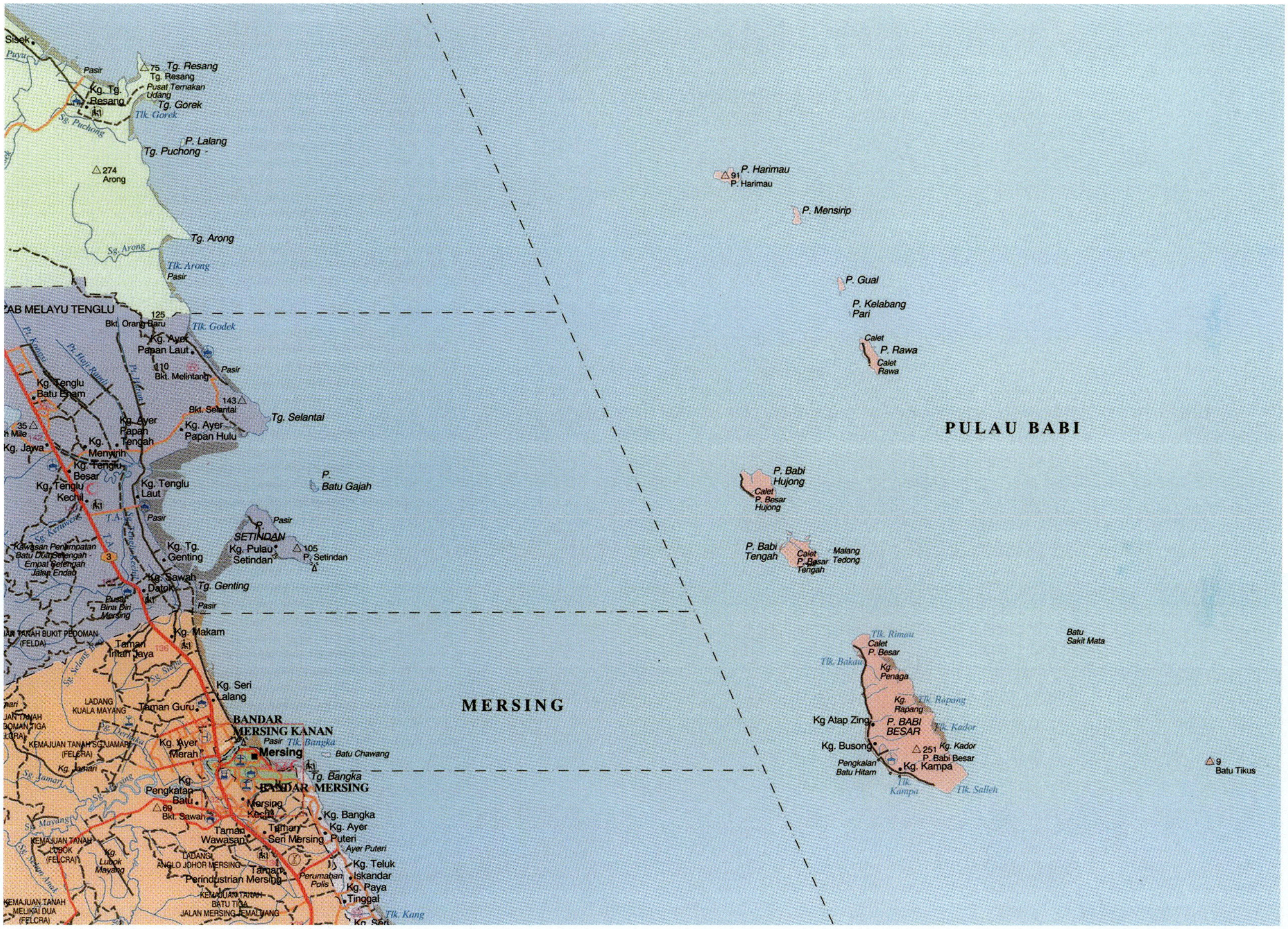
PULAU BABI
MERSING
P. Harimau
P. Mensirip
P. Gual
P. Kelabang
Pari
P. Rawa
P. Babi Hujong
P. Babi Tengah
Malang Tedong
P. BABI BESAR
Tlk. Rimau
Tlk. Bakau
Kg. Penaga
Kg. Rapang
Tlk. Rapang
Tlk. Kador
Kg Atap Zing
Kg. Busong
Kg. Kador
P. Babi Besar
Kg. Kampa
Pengkalan Batu Hitam
Tlk. Kampa
Tlk. Salleh
Batu Sakit Mata
Batu Tikus
Tg. Resang
Kg. Tg. Besang
Tg. Gorek
Tlk. Gorek
Sg. Puchong
P. Lalang
Tg. Puchong
Arong
Tg. Arong
Sg. Arong
Tlk. Arong
Tlk. Godek
Kg. Ayer Papan Laut
Bkt. Melintang
Bkt. Selantai
Tg. Selantai
Kg. Tenglu Batu Empam
Kg. Ayer Papan Tengah
Kg. Ayer Papan Hulu
Kg. Menyinin
Kg. Tenglu Besar
Kg. Tenglu Kechil
Kg. Tenglu Laut
Kg. Jawa
P. Batu Gajah
P. SETINDAN
Kg. Pulau Setindan
P. Setindan
Kg. Tg. Genting
Kg. Sawah Datok
Tg. Genting
Kg. Makam
Taman Intan Jaya
Kg. Seri Lalang
Taman Guru
BANDAR MERSING KANAN
Tlk. Bangka
Batu Chawang
Mersing
Kg. Ayer Merah
Tg. Bangka
BANDAR MERSING
Kg. Pengkalan Batu
Bkt. Sawah
Mersing Kechil
Kg. Bangka
Kg. Ayer Puteri
Taman Wawasan
Taman Seri Mersing
Kg. Teluk Iskandar
Kg. Paya Tinggal
Perindustrian Mersing
Tlk. Kang
Kg. Seri

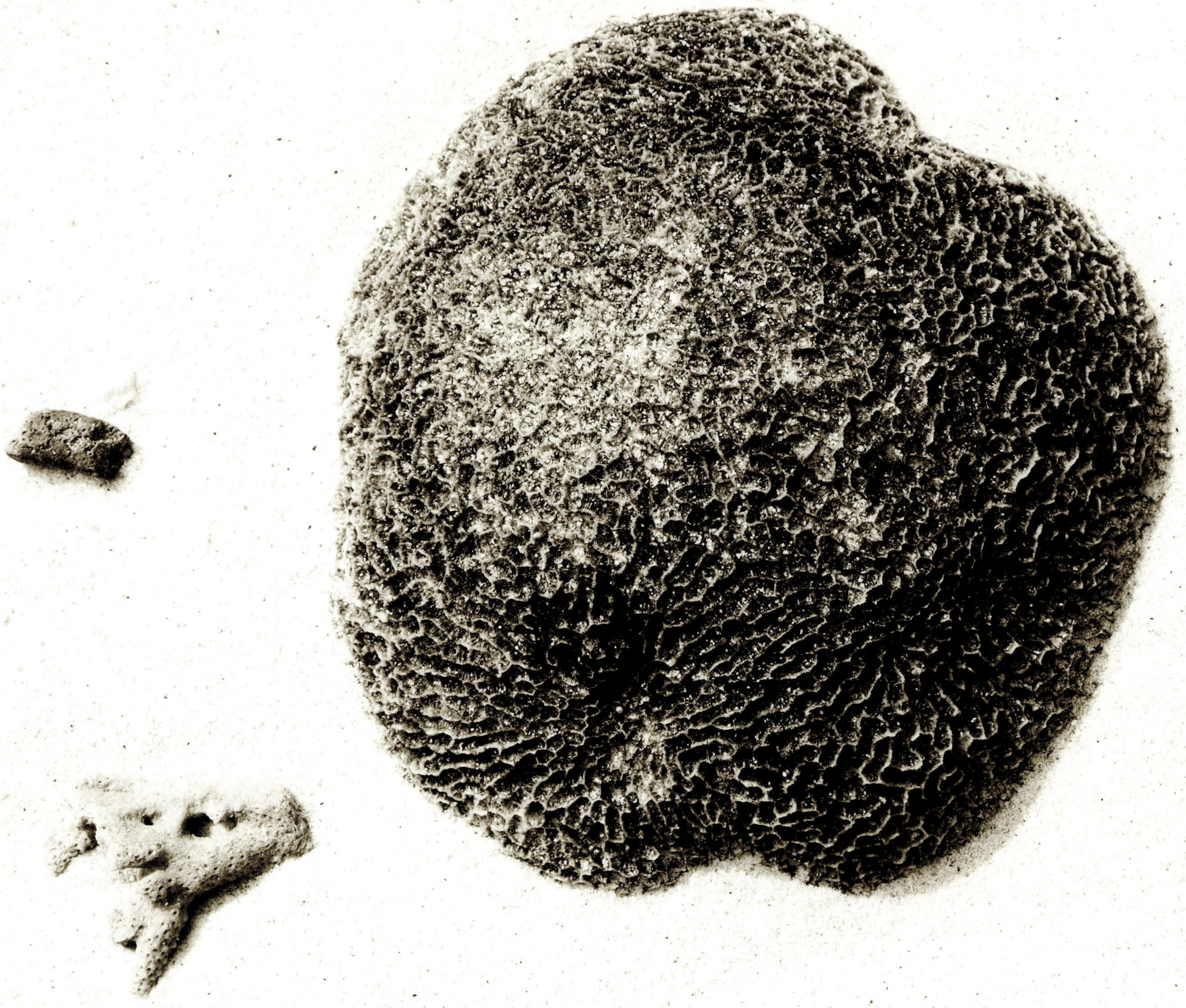

Corals on beach. *(Photo credit: Eric Frideen)*

Species list of Hard Coral, Fish, Marine Reptiles and Marine Mammals

HARD CORAL

Scientific Family Name	Scientific Species Name	Common Name
Holothuriidae	*Actinopygra lecanora*	Sea cucumber
Holothuriidae	*Bohadschia argus*	Sea cucumber
Holothuriidae	*Holothuria atra*	Sea cucumber
Holothuriidae	*Holothuria edulis*	Sea cucumber
Holothuriidae	*Holothuria leucospilata*	Sea cucumber
Holothuriidae	*Pearsonuthuria graeffei*	Sea cucumber
Stichopodidae	*Stichopus chloronotus*	Sea cucumber
Stichopodidae	*Stichopus ocellatus*	Sea cucumber
Stichopodidae	*Stichopus vastus*	Sea cucumber
Synaptidae	*Synaptula lamperti*	Sea cucumber
Diadematidae	*Diadema setosum*	Sea urchin
Diadematidae	*Diadema savignyi*	Sea urchin
Diadematidae	*Echinotrix calamaris*	Sea urchin
Diadematidae	*Echinotrix diadema*	Sea urchin
Echinometridae	*Parasalenia gratiosa*	Sea urchin
Toxopneustidae	*Tripneustes sp.*	Sea urchin
Acanthasteridae	*Acanthasther planci*	Crown-of-Thorns
Echinasteridae	*Echinaster luzonicus*	Sea star
Ophidiasteridae	*Fromia monilis*	Sea star
Ophidiasteridae	*Linckia laevigata*	Sea star
Ophidiasteridae	*Lickia multifora*	Sea star
Oreasterida	*Choriaster granulatus*	Sea star
Oreasterida	*Culcita novaguineae*	Sea star
Oreasterida	*Protoreaster nodosus*	Sea star
Comasteridae	*Comaster multifidus*	Featherstar
Comasteridae	*Comanthina sp.*	Featherstar
Stichodactylidae	*Heteractis magnifica*	Sea anemone
Stichodactylidae	*Heteractis crispa*	Sea anemone
Actinodendronidae	*Actinodendron sp.*	Sea anemone
Actiniidae	*Entacmaea quadricolor*	Sea anemone
Cerianthidae	*Cerianthus sp.*	Sea anemone
Phyllidiidae	*Phyllidia coelestis*	Nudibranch
Phyllidiidae	*Phyllidia elegans*	Nudibranch
Phyllidiidae	*Phyllidia varicosa*	Nudibranch
Chromodorididae	*Chromodoris reticulata*	Nudibranch
Chromodorididae	*Risbecia tryoni*	Nudibranch
Tridacnidae	*Tridacna crocea*	Giant clam
Tridacnidae	*Tridacna squamosa*	Giant clam
Tridacnidae	*Tridacna maxima*	Giant clam
Neritidae	*Nerita albicila*	Shell
Turbinidae	*Turbo petholatus*	Shell
Cypraeidae	*Cypraea tigris*	Shell

HARD CORAL

Scientific Family Name	Scientific Species Name	Common Name
Strombidae	*Lambis lambis*	Shell
Conidae	*Conus sp.*	Shell
Muricidae	*Drupella rugosa*	Shell
Spondylidae	*Spondylus tenebrosus*	Shell
Ostereidae	*Lopha cristagalli*	Shell
Malleidae	*Malleus albus*	Shell
Pectinidae	*Pedum spondyloideum*	Shell
Virgulariidae	*Virgularia sp.*	Sea pen

FISH

Scientific Family Name	Scientific Species Name	Common Name
Acanthuridae	*Acanthurus lineatus*	Striped Surgeonfish
Apogonidae	*Cheilodipterus macrodon*	Tiger Cardinalfish
Atherinidae	*Hypoatherina bamesi*	Slender Silverside
Balistidae	*Balistoides viridescens*	Titan Triggerfish
Balistidae	*Pseudobalistes flavimarginatus*	Yellowmargin Triggerfish
Batrachoididae	*Halophyme diemensis*	Banded Toadfish
Belonidae	*Strongylura incisa*	Reef Needlefish
Bothidae	*Bothus mancus*	Peacock Flounder
Caesionidae	*Pterocaesio tile*	Bluestreak Fusilier
Caesionidae	*Caesio xanthonota*	Yellowback Fusilier
Caesionidae	*Caesio teres*	Blue and yellow Fusilier
Callionymidae	*Dactylopus kuiteri*	Orange and Black Dragonet
Carangidae	*Alectis ciliaris*	African Pompano
Carangidae	*Carangoides orthogrammus*	Yellowspotted Trevally
Carangidae	*Carangoides fulvoguttatus*	Goldspotted Trevally
Carangidae	*Scomberoides commersonnianus*	Talang Queenfish
Carangidae	*Trachinotus baillonii*	Small-spotted Dart
Carangidae	*Gnathanodon speciosus*	Golden Trevally
Carangidae	*Trachinotus blochii*	Silver Pompano
Carcharhinidae	*Carcharhinus melanopterus*	Blacktip Reef Shark
Centriscidae	*Aeoliscus strigatus*	Razorfish
Chaetodontidae	*Chaetodon adiergastos*	Panda Butterflyfish
Chaetodontidae	*Chaetodon octofasciatus*	Eight-banded Butterflyfish
Chaetodontidae	*Heniochus acuminatus*	Longfin Bannerfish
Chaetodontidae	*Heniochus varius*	Humphead Bannerfish
Chaetodontidae	*Parachaetodon ocellatus*	Ocellated Coralfish
Chaetodontidae	*Chelmon rostratus*	Long-beaked Coralfish
Chanidae	*Chanos chanos*	Milkfish
Congridae	*Heteroconger hassi*	Spotted Garden Eel
Dasyatidae	*Neotrygon kuhlii*	Blue-spotted Stingray

FISH Scientific Family Name	Scientific Species Name	Common Name
Dasyatidae	*Taeniura lymma*	Bluespotted Ribbontail ray
Diodontidae	*Cyclichthys spilostylus*	Yellowspotted Burrfish
Diodontidae	*Diodon holocanthus*	Balloonfish
Diploprionini	*Diploprion bifascaitum*	Double-banded Soapfish
Echeneidae	*Echensis naucrates*	Sharksucker
Ephippidae	*Platax teira*	Longfin Spadefish
Ephippidae	*Platax batavianus*	Batavia Spadefish
Ephippidae	*Platax orbicularis*	Circular Spadefish
Ephippidae	*Platax pinnatus*	Pinnate Spadefish
Fistulariidae	*Fistularia commersonii*	Cornetfish
Gobiesocidae	*Diademichthys lineatus*	Urchin Clingfish
Gobiidae	*Amblyeleotris randalli*	Randalls' Shrimpgoby
Gobiidae	*Amblyeleotris steinitzi*	Steinitz' Shrimpgoby
Gobiidae	*Cryptocentrus fasciatus*	Barred Shrimpgoby
Haemulidae	*Plectorhinchus vittatis*	Oriental Sweetlips
Haemulidae	*Diagramma pictum*	Silver Sweetlips
Hypnidae	*Hypnos monopterygium*	Coffin Ray
Labridae	*Epibulus insidiator*	Slingjaw Wrasse
Labridae	*Hemigymnus melapterus*	Blackeye Thicklip
Labridae	*Choerodon oligacanthus*	Tuskfish
Labridae	*Choerodon anchorago*	Anchor Tuskfish
Labridae	*Labroides bicolor*	Bicolour Cleaner Wrasse
Labridae	*Thalassoma lunare*	Crescent wrasse
Lutjanidae	*Lutjanus ehrenbergii*	Blackspot Snapper
Lutjanidae	*Lutjanus kasmira*	Bluestripe Snapper
Malacanthidae	*Malacanthus latovittatus*	Blue Blanquillo
Megalopidae	*Megalops cyprinoides*	Indo pacific Tarpon
Monocanthidae	*Aluterus scriptus*	Scrawled Filefish
Monocanthidae	*Chaetodermis penicilligera*	Leafy Filefish
Monocanthidae	*Pseudomonacanthus macrurus*	Strapweed Filefish
Monodactylidae	*Monodactylus argenteus*	Diamond Fish
Mugilidae	*Crenimugil crenilabis*	Fringelip Mullet
Mulliade	*Parupeneus indicus*	Indian Goatfish
Mulliade	*Upeneus tragula*	Freckled Goatfish
Muraenidae	*Gymnothorax javanicus*	Giant Moray
Muraenidae	*Gymnothorax fimbriatus*	Fimbriated Moray
Muraenidae	*Gymnothorax isingteena*	Spotted Moray
Muraenidae	*Gymnothorax thyrsoideus*	White Eyed Moray
Nemipteridae	*Scolopsis bilineatus*	Bridled Monocle Bream
Nemipteridae	*Scolopsis torquatus*	Whitecheek Monocle Bream
Ophichthidae	*Callechelys marmorata*	Marbled Snake Eel
Ophichthidae	*Ophichthus melanochir*	Black Finned Snake Eel
Ostraciidae	*Ostracion rhynohynchus*	Shortnose Boxfish
Pegasidae	*Pegasus volitans*	Slender Seamoth

FISH Scientific Family Name	Scientific Species Name	Common Name
Pempheridae	*Pempheris oualensis*	Copper Sweeper
Pinguipedidae	*Parapercis hexophthalma*	Speckled Sandperch
Platycephalidae	*Cymbacephalus beauforti*	Crocodile Flathead
Platycephalidae	*Thysanophrys arenicola*	Broadhead Flathead
Platycephalidae	*Thysanophrys carbunculus*	Horned Flathead
Plotosidae	*Paraplotosus albilabris*	Whitelipped Catfish
Pomacanthidae	*Pomacanthus annularis*	Blue-ringed Angelfish
Pomacanthidae	*Pomacanthus sexstriatus*	Six-banded Angelfish
Pomacanthidae	*Chaetodontoplus mesoleucus*	Vermiculated Angelfish
Pomacanthidae	*Apolemichthys trimaculatus*	Threespot Angelfish
Pomacentridae	*Abudefduf septemfasciatus*	Banded Sergeant
Pomacentridae	*Amphiprion perideraion*	Pink Anemonefish
Pomacentridae	*Amphiprion clarkii*	Clarks' Anemonefish
Pomacentridae	*Amphiprion frenatus*	Tomato Anemonefish
Pomacentridae	*Amphiprion polymnus*	Saddleback Anemonefish
Pomacentridae	*Amphiprion ocellaris*	False Clown Anemonefish
Pomacentridae	*Neoglyphidodon melas*	Black Damsel
Pomacentridae	*Dascyllus trimaculatus*	Threespot Dascyllus
Pomacentridae	*Dischistodus chrysopoecilus*	Palespot Damsel
Pomacentridae	*Amblyglyphidodon aureus*	Golden damsel
Ptereleotridae	*Nemateleotris helfrichi*	Helfrichs' Dartfish
Rachycentridae	*Rachycentron canadum*	Cobia
Rhinobatidae	*Rhinobatus typus*	Shovelnose Ray
Scaridae	*Bolbometopon muricatum*	Bumphead Parrotfish
Scaridae	*Chlorurus sordidus*	Bullethead Parrotfish
Scaridae	*Scarus ghobban*	Blue barred Parrotfish
Scorpaenidae	*Dendrochirus biocellatus*	Twinspot Lionfish
Scorpaenidae	*Scorpaenopsis diabolus*	Devil Scorpionfish
Scorpaenidae	*Scorpaenopsis macrochir*	Flasher Scorpionfish
Scorpaenidae	*Pterois volitans*	Common Lionfish
Scorpaenidae	*Dendrochirus zebra*	Zebra Lionfish
Serranidae	*Anyperodon leucogrammicus*	Slender Grouper
Serranidae	*Cephalopholis cyanostigma*	Bluespotted Grouper
Serranidae	*Cephalopholis argus*	Peacock Grouper
Serranidae	*Cephalopholis miniata*	Coral Grouper
Serranidae	*Cromileptes altivelis*	Barramundi
Serranidae	*Epinephelus corallicola*	Coral Rock Grouper
Serranidae	*Epinephelus areolatus*	Areolate Grouper
Serranidae	*Epinephelus lanceolatus*	Giant Grouper
Serranidae	*Plectropomus oligacanthus*	Highfin Coral Grouper
Sigaidae	*Siganus vulpinus*	Foxface Rabbitfish
Sigaidae	*Siganus guttatus*	Golden Rabbitfish
Sigaidae	*Siganus javus*	Java Rabbitfish
Sphyraenidae	*Sphyraena barracuda*	Great Barracuda

FISH

Scientific Family Name	Scientific Species Name	Common Name
Sphyraenidae	*Sphyraena jello*	Pickhandle Barracuda
Sphyraenidae	*Sphyraena qenie*	Blackfin Barracuda
Sphyraenidae	*Sphyraena obtusata*	Yellowtail Barracuda
Synanceiidae	*Inimicus didactylus*	Spiny Devilfish
Synanceiidae	*Synanceia horrida*	Estuarine Stonefish
Syngnathidae	*Micrognathus andersonii*	Shortnose Pipefish
Syngnathidae	*Trachyrhamphus bicoarctatus*	Short Tailed Pipefish
Syngnathidae	*Hippichthys cyanospilus*	Estuary Pipefish
Synodontidae	*Synodus dermatogenys*	Clearfin Lizardfish
Synodontidae	*Synodus variegatus*	Reef Lizardfish
Tetraodontidae	*Arothron caeruleopunctatus*	Bluespotted Puffer
Tetraodontidae	*Arothron stellatus*	Star Puffer
Tetraodontidae	*Arothron mappa*	Map Puffer
Zanclidae	*Zanclus comutus*	Moorish Idol

MARINE REPTILES

Scientific Family Name	Scientific Species Name	Common Name
Cheloniidae	*Chelonia mydas*	Green Turtle
Cheloniidae	*Eretmochelys imbricata*	Hawksbill Turtle

MARINE MAMMALS

Scientific Family Name	Scientific Species Name	Common Name
Delphinidae	*Sousa chinensis*	Indo-Pacific Humpback Dolphin
Delphinidae	*Tursiops aduncus*	Indo-Pacific Bottlenose dolphin
Delphinidae	*Delphinus capensis*	Long-beaked Common Dolphin
Dugongidae	*Dugong dugon*	Dugong
Phocoenidae	*Neophocaena phocaenoides*	Indo-Pacific Finless Porpoise

INVERTEBRATES

Scientific Family Name	Scientific Species Name	Common Name
Holothuriidae	*Actinopygra lecanora*	Sea cucumber
Holothuriidae	*Bohadschia argus*	Sea cucumber
Holothuriidae	*Holothuria atra*	Sea cucumber
Holothuriidae	*Holothuria edulis*	Sea cucumber
Holothuriidae	*Holothuria leucospilata*	Sea cucumber
Holothuriidae	*Pearsonuthuria graeffei*	Sea cucumber

INVERTEBRATES Scientific Family Name	Scientific Species Name	Common Name
Stichopodidae	*Stichopus chloronotus*	Sea cucumber
Stichopodidae	*Stichopus ocellatus*	Sea cucumber
Stichopodidae	*Stichopus vastus*	Sea cucumber
Synaptidae	*Synaptula lamperti*	Sea cucumber
Diadematidae	*Diadema setosum*	Sea urchin
Diadematidae	*Diadema savignyi*	Sea urchin
Diadematidae	*Echinotrix calamaris*	Sea urchin
Diadematidae	*Echinotrix diadema*	Sea urchin
Echinometridae	*Parasalenia gratiosa*	Sea urchin
Toxopneustidae	*Tripneustes sp.*	Sea urchin
Acanthasteridae	*Acanthasther planci*	Crown-of-Thorns
Echinasteridae	*Echinaster luzonicus*	Sea star
Ophidiasteridae	*Fromia monilis*	Sea star
Ophidiasteridae	*Linckia laevigata*	Sea star
Ophidiasteridae	*Lickia multifora*	Sea star
Oreasterida	*Choriaster granulatus*	Sea star
Oreasterida	*Culcita novaguineae*	Sea star
Oreasterida	*Protoreaster nodosus*	Sea star
Comasteridae	*Comaster multifidus*	Featherstar
Comasteridae	*Comanthina sp.*	Featherstar
Stichodactylidae	*Heteractis magnifica*	Sea anemone
Stichodactylidae	*Heteractis crispa*	Sea anemone
Actinodendronidae	*Actinodendron sp.*	Sea anemone
Actiniidae	*Entacmaea quadricolor*	Sea anemone
Cerianthidae	*Cerianthus sp.*	Sea anemone
Phyllidiidae	*Phyllidia coelestis*	Nudibranch
Phyllidiidae	*Phyllidia elegans*	Nudibranch
Phyllidiidae	*Phyllidia varicosa*	Nudibranch
Chromodorididae	*Chromodoris reticulata*	Nudibranch
Chromodorididae	*Risbecia tryoni*	Nudibranch
Tridacnidae	*Tridacna crocea*	Giant clam
Tridacnidae	*Tridacna squamosa*	Giant clam
Tridacnidae	*Tridacna maxima*	Giant clam
Neritidae	*Nerita albicila*	Shell
Turbinidae	*Turbo petholatus*	Shell
Cypraeidae	*Cypraea tigris*	Shell
Strombidae	*Lambis lambis*	Shell
Conidae	*Conus sp.*	Shell
Muricidae	*Drupella rugosa*	Shell
Spondylidae	*Spondylus tenebrosus*	Shell
Ostereidae	*Lopha cristagalli*	Shell
Malleidae	*Malleus albus*	Shell
Pectinidae	*Pedum spondyloideum*	Shell
Virgulariidae	*Virgularia sp.*	Sea pen

Species List of Birds

BIRDS Species Name	Common Name	Habitat Encountered	Status	Area
Bulweria bulwerii	Bulwer's Petrel	Pelagic	Migrant	Offshore
Oceanodroma monorhis	Swinhoe's Storm Petrel	Pelagic	Migrant	Offshore
Fregata andrewsii	Christmas Island Frigatebird	Pelagic	Non-breeding visitor	Offshore
Fregata ariel	Lesser Frigatebird	Pelagic	Non-breeding visitor	Pulau Tengah, Pulau Pemanggil, offshore
Sula leucogaster	Brown Booby	Pelagic	Migrant	Offshore
Ixobrychus sinensis	Yellow Bittern	Coastal scrub	Migrant	Pulau Tengah, Pulau Pemanggil
Ixobrychus flavicollis	Black Bittern	Swamp	Scarce migrant	Pulau Tengah
Ardea purpurea	Purple Heron	Estuary	Migrant	Mersing
Ardea alba	Great Egret	Beach	Migrant	Pulau Tengah
Egretta eulophotes	Chinese Egret	Estuary	Migrant	Mersing
Egretta garzetta	Little Egret	Mudflat, beach	Resident/winter visitor	Mersing
Egretta sacra	Pacific Reef Egret	Beach, rocky outcrop	Resident	Pulau Tengah, Pulau Pemanggil
Bubulcus ibis	Cattle Egret	Coastal scrub	Migrant	Pulau Tengah, Pulau Pemanggil
Butorides striata	Striated Heron	Beach, rocky outcrop	Resident	Pulau Tengah, Pulau Tinggi
Nycticorax nycticorax	Black-crowned Night- Heron	Beach, rocky outcrop	Migrant	Pulau Tengah
Paidon haliaetus	Osprey	Sea	Migrant	Mersing / Pulau Tengah
Pernis ptilorhynchus	Oriental Honey-Buzzard	Open are	Passage migrant	Pulau Tengah
Lophotriorchis kienerii	Rufous-bellied Eagle	Beach	Migrant	Mersing
Butastur indicus	Grey-faced Buzzard	Open area	Migrant	Pulau Tengah
Accipiter soloensis	Chinese Sparrowhawk	Open area	Migrant	Pulau Tengah
Accipiter gularis	Japanese Sparrowhawk	Open area	Migrant	Pulau Tengah
Haliastur indus	Brahminy Kite	Open area, coastal shrub	Resident	Pulau Tengah, Mersing, Pulau Aur, Pulau Tinggi, Pulau Pemanggil
Haliaaetus leucogaster	White-bellied Sea-Eagle	Open area, beach	Resident	Pulau Tengah, Mersing, Pulau Aur, Pulau Pemanggil, Pulau Tioman
Amaurornis phoenicurus	White-breasted Waterhen	Swamp	Resident	Mersing
Pluvialis fulva	Pacific Golden-Plover	Beach, mudflat	Migrant	Mersing
Vanelllus cineerus	Red-wattled Lapwing	Long beach	Migrant	Pulau Tengah
Charadrius mongolus	Lesser Sand-Plover	Beach, mudflat	Migrant	Pulau Tengah, Mersing
Charadrius leschenaultii	Greater Sand-Plover	Beach, mudflat	Migrant	Mersing
Charadrius peronii	Malaysian Plover	Beach, mudflat	Rare resident	Mersing
Charadrius alexandrius	Kentish Plover	Beach, mudflat	Winter visitor	Mersing
Xenus cinereus	Terek Sandpiper	Beach, mudflat	Migrant	Mersing
Actitis hypoleucos	Common Sandpiper	Beach, rocky outcrop	Migrant	Pulau Tengah, Mersing, Pulau Pemanggil
Tringa brevipes	Grey-tailed Tattler	Beach, mudflat	Migrant	Mersing
Tringa nebularia	Common Greenshank	Beach, mudflat	Migrant	Mersing
Tringa guttifer	Nordmann's Greenshank	Estuary	Vagrant	Mersing
Tringa stagnatilis	Marsh Sandpiper	Estuary	Rare migrant	Mersing
Tringa totanus	Common Redshank	Beach, mudflat	Migrant	Mersing
Numenius phaeopus	Whimbrel	Beach , mudflat	Migrant	Pulau Tengah, Mersing
Numenius madagascariensis	Far Eastern Curlew	Estuary	Vagrant	Mersing
Limosa lapponica	Bar tailed Godwit	Estuary	Migrant	Mersing
Arenaria interpres	Ruddy Turnstone	Beach, mudflat	Migrant	Mersing
Calidris tenuirostris	Great Knot	Estuary	Migrant	Mersing
Calidris canutus	Red Knot	Estuary	Migrant	Mersing

BIRDS

Species Name	Common Name	Habitat Encountered	Status	Area
Calidris falcinellus	Broad-billed Sandpiper	Estuary	Migrant	Mersing
Calidris ferruginea	Curlew Sandpiper	Estuary	Migrant	Mersing
Calidris ruficolis	Red-necked Stint	Beach	Migrant	Mersing
Calidris alba	Sanderling	Beach, mudflat	Migrant	Mersing
Gallinago gallinago	Common Snipe	Swamp	Migrant	Pulau Tengah
Phalaropus lobatus	Red-necked Phalarope	Pelagic	Rare migrant	Offshore Mersing
Glareola maldivarum	Oriental Pratincole	Coastal rocks	Migrant	Pulau Besar
Stercorarius pomarinus	Pomarine Jaeger	Pelagic	Migrant	Offshore Mersing
Stercorarius longicaudus	Long-tailed Jaeger	Pelagic	Migrant	Offshore Mersing
Stercorarius parasiticus	Parasitic Jaeger	Pelagic	Migrant	Offshore
Chroicocephalus ridibundus	Black-headed Gull	Beach	Migrant	Mersing Estuary
Onychoprion fuscatus	Sooty Tern	Pelagic	Vagrant	Offshore
Onychoprion anaethus	Bridled Tern	Pelagic	Migrant	Offshore
Sternula albifrons	Little Tern	Pelagic, coastal	Migrant	Mersing. offshore
Gelochelidon nilotica	Gull-billed Tern	Pelagic	Rare migrant	Mersing, offshore
Hydroprogne caspia	Caspian Tern	Estuary	Vagrant	Mersing
Chilidonias leucopterus	White-winged Tern	Pelagic, coastal	Migrant	Offshore, Mersing
Sterna dougallii	Roseate Tern	Pelagic	Migrant	Offshore
Sterna sumatrana	Black-naped Tern	Island, pelagic	Migrant summer Resident breeding	Pulau Tengah
Sterna hirundo	Common Tern	Pelagic	Rare migrant	Mersing. offshore
Thalasseus bergii	Great Crested Tern	Beach, pelagic	Winter visitor	Mersing, offshore
Thalassesseus bengalensis	Lesser Crested Tern	Sea, estuary	Rare passage migrant	Mersing Estuary, pelagic
Columba livia	Rock Pigeon	Garden	Feral resident	Mersing
Streptopelia chinensis	Spotted Dove	Garden, open area, coastal shrub	Resident	Pulau Tengah, Mersing, Pulau Tinggi
Geopelia striata	Zebra Dove	Beach, garden	Resident	Mersing, Pulau Tinggi
Treron vernans	Pink-necked Pigeon	Forest, coastal shrub, garden	Resident	Pulau Tengah, Mersing, Pulau Pemanggil
Caloenas nicobarica	Nicobar Pigeon	Secondary forest	Scarce Visitor	Pulau Tengah
Ducula bicolor	Pied Imperial-Pigeon	Forest, coastal shrub	Resident	Pulau Tengah, Pulau Pemanggil
Cacomantis merulinus	Plaintive Cuckoo	Coastal shrub, forest	Resident	Pulau Tengah
Surniculus sp.	Drongo Cuckoo sp.	Secondary forest	Migrant	Pulau Tengah
Eudynamys scolopaceus	Asian Koel	Garden	Migrant	Pulau Tengah, Mersing
Centropus bengalensis	Lesser Coucal	Coastal shrub	Resident	Pulau Tengah
Caprimulgus affinis	Savanna Nightjar	Secondary forest	Migrant	Pulau Tengah
Caprimulgus macrurus	Large-tailed Nightjar	Secondary forest	Migrant/summer resident	Pulau Tengah
Aerodramus maximus	Black-nest Swiftlet	Open area	Resident	Mersing
Aerodramus germani	Germain's Swiftlet	Open area, beach	Resident	Pulau Tengah, Mersing
Apus pacificus	Pacific Swift	Open area	Passage migrant	Pulau Tengah
Hirundapus giganteus	Brown-backed Needletail	Open Area	Passage migrant	Pulau Tengah
Apus affinis	House Swift	Open area	Resident	Pulau Tengah
Alcedo atthis	Common Kingfisher	Open area	Passage migrant	Pulau Tengah
Pelargopsis capensis	Stork-billed Kingfisher	Mangrove	Resident	Mersing
Halcyon smyrnensis	White-throated Kingfisher	Secondary forest	Resident	Mersing
Todiramphus chloris	Collared Kingfisher	Coastal shrub	Resident	Pulau Tengah, Pulau Tinggi, Mersing
Merops philippinus	Blue-tailed Bee-eater	Beach	Passage migrant	Pulau Tengah, Mersing
Coracias benghalensis	Dollarbird	Open area, coastal shrub	Resident	Pulau Tengah, Pulau Tinggi

BIRDS

Species Name	Common Name	Habitat Encountered	Status	Area
Upupa epops	Eurasian Hoopoe	Open area	Vagrant	Pulau Tengah
Anthracoceros albirostris	Oriental Pied-Hornbill	Woodland/open areas	Resident	Mersing
Psilopogon lineatus	Lineated Barbet	Woodland	Resident	Mersing
Falco peregrinus	Peregrine Falcon	Coastal scrub	Migrant	Pulau Tengah, Mersing
Pachycephala cinerea	Mangrove Whistler	Coastal scrub	Resident	Pulau Tengah
Lanius criscatus	Brown Shrike	Coastal scrub	Migrant	Pulau Tengah
Oriolus chinensis	Black-naped Oriole	Coastal scrub	Resident	Pulau Tengah, Mersing
Dicrurus macrocercus	Black Drongo	Open area	Migrant	Pulau Tengah
Hypothymis azurea	Black naped Monarch	Forest	Resident	Pulau Tengah
Corvus macrorhynchos	Large-billed Crow	Open area, coastal shrub	Resident	Pulau Tengah, Mersing, Pulau Aur, Pulau Tinggi, Pulau Pemanggil
Hirundo rustica	Barn Swallow	Open area, beach, rock	Migrant/Summer resident	Pulau Tengah, Mersing, Pulau Pemanggil, Pulau Aur
Hirundo tahitica	Pacific Swallow	Open area, beach, rock	Resident	Pulau Tengah, Mersing, Pulau Pemanggil
Pycnonotus goiavier	Yellow-vented Bulbul	Garden, coastal shrub	Resident	Mersing
Pycnonotus plumosus	Olive-winged Bulbul	Forest, coastal shrub, garden	Resident	Pulau Tengah, Pulau Aur, Pulau Tinggi
Phylloscopus borealis	Arctic Warbler	Secondary forest, coastal shrub	Migrant	Pulau Tengah
Orthotomus ruficeps	Ashy Tailorbird	Coastal shrub	Resident	Pulau Tengah, Mersing
Macrorus gularis	Pin-striped Tit Babbler	Forest	Resident	Pulau Tengah
Malacocinla abbotti	Abbott's Babbler	Forest	Resident	Pulau Tengah
Muscicapa latirostris	Asian Brown Flycatcher	Coastal shrub	Migrant	Pulau Tengah
Copsychus saularis	Oriental Magpie-Robin	Garden, coastal shrub	Resident	Pulau Tengah, Mersing
Copsychus malabaricus	White-rumped Shama	Forest	Resident	Pulau Tengah
Cyanoptila cyanomelana/ cumatilis	Blue and White/ Zappey's Flycatcher	Forest	Migrant	Pulau Tengah
Ficedula zanthopygia	Korean Flycatcher	Woodland	Migrant	Pulau Tengah
Ficedula mugimaki	Mugimaki Flycatcher	Forest	Migrant	Pulau Tengah
Aplonis panayensis	Asian Glossy Starling	Garden, beach	Resident	Mersing, Pulau Aur, Pulau Tinggi
Gracula religiosa	Common Hill Myna	Forest	Resident	Pulau Tinggi, Pulau Pemanggil
Acridotheres javanicus	Javan Myna	Garden, coastal shrub	Feral resident	Mersing, Pulau Pemanggil
Acridotheres tristis	Common Myna	Garden, coastal shrub	Resident	Pulau Tengah, Mersing, Pulau Pemanggil
Sturnia sturnuna	Daurian Starling	Coastal shrub	Migrant	Pulau Tengah
Dicaeum cruentatum	Scarlet-backed Flowerpecker	Coastal shrub, forest	Resident	Pulau Tengah, Mersing
Anthreptes malacensis	Plain-throated Sunbird	Garden, coastal shrub	Resident	Pulau Tengah, Mersing, Pulau Tinggi
Leptocoma brasiliana	Van Hasselt's Sunbird	Coastal shrub, garden	Resident	Pulau Tengah
Cinnyris jugularis	Olive-backed Sunbird	Coastal shrub, garden	Resident	Pulau Tengah, Mersing
Motacilla tschutchensis	Eastern Yellow Wagtail	Beach	Migrant	Pulau Tengah, Mersing
Anthus rufulus	Oriental Pipit	Beach	Resident	Mersing
Passer montanus	Eurasian Tree Sparrow	Urban	Resident	Mersing, Pulau Aur, Pulau Tinggi
Lonchura punctulata	Scaly-breasted Munia	Woodland	Resident	Mersing

Botany Species List

Previous records ("PR") are indicated for Pulau Tengah: C, Corner (1985), Q, Quijrin (2012, unpublished). All specimens cited ("Specimen n°") are collected by B. Senterre (6595-6645), B. Senterre & K.F. Senafi (6646-6682) and B. Senterre & A. Ahmad (6683-6902). Observation records are indicated with the abbreviation "o.r.", and when nothing is indicated for the "Specimen n°" the species record is from the literature only. Observed characters ("Compo") are abbreviated as follows: st, sterile; fert, fertile; flb, flower bud; fl, flower; fr, fruit. Prefered habitat-types ("Hab.-Type") are abbreviated as follows: 2F, secondary forests; An, anthropic areas; CF, coastal forests; Co, coastal sandy beaches; CR, coastal rocky shores; CS, coastal marshes/swamps; DF, dry littoral forests; MF, mesophilous littoral forests; MF, mangrove forest; Pi, pioneer vegetation; VF, valley littoral forests. Biological Types (TB) are abbreviated as follows: T, trees; S, shrubs; L, Lianas; V, vines; H, terrestrial herbs; E, epiphytes (incl. rupicolous species). Origin status (Orig) is abbreviated as follows: end, endemic; ind, indigenous; cult, cultivated; nat, naturalized; inv, invasive (see definitions in materials and methods). Rarity is abbreviated as follows: R, rare; O, occasional; F, frequent; C, common; A, abundant.

Dicotyledons (Palaeodicots and Eudicots)

Name	Family Species	PR	Specimen n°	Compo	Hab.-Type	TB	Orig	Rarity
Acanthaceae	*Asystasia gangetica (L.) T. Anderson subsp. micrantha (Nees) Ensermu*		6775	fl, fr	An	H	inv	F
Acanthaceae	*Staurogyne setigera (Nees) Kuntze*		6605	fl	VF	H	ind	R
Aizoaceae	*Sesuvium portulacastrum (L.) L.*		6847	fert	An	H	ind	F
Amaranthaceae	*Alternanthera sessilis (L.) R. Br. ex Roth*		6896	fert	An	H	cult	O
Amaranthaceae	*Amaranthus dubius Mart. ex Thell.*		6845	fert	An	H	nat	F
Anacardiaceae	*Anacardium occidentale L.*		6744	st	2F	T	nat	O
Anacardiaceae	*Buchanania arborescens (Blume) Blume*		6610, 6632, 6717	st	MF	T	ind	F
Anacardiaceae	*Mangifera indica L.*		o.r.		An	T	cult	F
Anacardiaceae	*Mangifera pentandra Hook.f.*		6829	st	CF	T	ind	O
Annonaceae	*Desmos dasymaschalus (Blume) Safford var. wallichii (Hook.f. & Thomson) Ridl.*		6616	st	MF	S	ind	F
Apocynaceae	*Allamanda cathartica L.*		o.r.		An	V	cult	F
Apocynaceae	*Alstonia spatulata Blume*		6699	st	MF	T	ind	R
Apocynaceae	*Alyxia reinwardtii Blume*		6641, 6748	st	MF	V	ind	O
Apocynaceae	*Catharanthus roseus (L.) G. Don*		o.r.		An	H	nat	F
Apocynaceae	*Cerbera manghas L.*		6761	st	CF	T	ind	R
Apocynaceae	*Cynanchum ovalifolium Wight*		6745	fl, fr	Pi	V	ind	O
Apocynaceae	*Dischidia major (Vahl) Merr.*	C	o.r.		DF, CF	V	ind	O
Apocynaceae	*Hoya diversifolia Blume*	C	6899	fl	CF	V	ind	O
Apocynaceae	*Nerium oleander L.*		o.r.		An	S	cult	O
Apocynaceae	*Parsonsia alboflavescens (Dennst.) Mabb.*		6825	fl	Co	V	ind	R
Apocynaceae	*Plumeria obtusa L.*		o.r.		An	T	cult	F
Araliaceae	*Arthrophyllum diversifolium Blume*		6794	st	MF	T	ind	F
Asteraceae	*Ageratum conyzoides L.*		6849	fl	An	H	nat	F
Asteraceae	*Blumea balsamifera (L.) DC.*		6872	st	Pi	S	ind?	R
Asteraceae	*Chromolaena odorata (L.) R.M. King. & H. Robuns.*		6767	fl	An, Pi	H	inv	C
Asteraceae	*Emilia sonchifolia (L.) DC.*		6838	fl, fr	An	H	ind?	F
Asteraceae	*Mikania cordata (Burm.f.) Robinson*		6685	fl	CS	V	ind	F
Asteraceae	*Synedrella nodiflora Gaertn.*		6776	fl, fr	An	H	nat	F
Asteraceae	*Tridax procumbens L.*		6837	fl, fr	An	H	nat	F
Asteraceae	*Vernonia cinerea (L.) Less.*		6836	fl, fr	An	H	ind?	F
Asteraceae	*Wollastonia biflora (L.) DC.*	C	6805 st		An	H	ind?	R
Boraginaceae	*Cordia subcordata Lam.*	C			CF	T	ind	
Boraginaceae	*Tournefortia argentea L.f.*	C			Co	S	ind	
Burseraceae	*Canarium pilosum Benn.*		6795	st	VF	T	ind	F

Dicotyledons (Palaeodicots and Eudicots)

Name	Family Species	PR	Specimen n°	Compo	Hab.-Type	TB	Orig	Rarity
Capparaceae	*Capparis micracantha DC. subsp. korthalsiana (Miq.) M. Jacobs*		6638	st	MF	S	ind	R
Caricaceae	*Carica papaya L.*		o.r.		An	T	cult	F
Casuarinaceae	*Casuarina equisetifolia L.*	C	6648	fr	An	T	ind	F
Celastraceae	*Salacia sp.1*		6624	st	MF	L	ind	F
Chrysobalanaceae	*Maranthes corymbosa Blume*		6702, 6714, 6732, 6758	st	MF	T	ind	F
Cleomaceae	*Arivela viscosa (L.) Raf. var. viscosa*		6848	fl, fr	An	H	nat	F
Clusiaceae	*Calophyllum depressinervosum M.R. Hend. & Wyatt-Sm.*		6614	st	MF	T	ind	O
Clusiaceae	*Calophyllum inophyllum L.*		6735, 6756	st	MF, CF	T	ind	F
Clusiaceae	*Clusia rosea L.*		6831	fl	An	S	cult	O
Clusiaceae	*Garcinia mangostana L*	C	o.r.		An	T	cult	F
Clusiaceae	*Garcinia nigrolineata Planch. ex T. Anderson*		6609, 6757	fr	MF, DF	T	ind	F
Clusiaceae	*Garcinia parvifolia (Miq.) Miq.*		6801	st	MF	S	ind	F
Combretaceae	*Terminalia catappa L.*	C	6655	st	CF	T	ind	F
Connaraceae	*Connarus monocarpus L. subsp. malayensis Leenh.*		6859	st	MF	L	ind	O
Connaraceae	*Rourea fulgens Planch.*		6626	st	MF	L	ind	F
Connaraceae	*Rourea minor (Gaertn.) Alston*		6619	st	MF	L	ind	F
Connaraceae	*Rourea rugosa Planch.*		6612	st	MF	L	end	F
Convolvulaceae	*Ipomoea littoralis Blume*		6654, 6686	fl	CF	V	ind	O
Convolvulaceae	*Ipomoea pes-caprae (L.) Sweet*	Q	6807	fl	Co	V	ind	C
Cucurbitaceae	*Melothria pendula L.*		6785	fl, fr	An	V	nat	F
Dilleniaceae	*Dillenia pulchella (Jack) Gilg*		6857	st	DF	T	ind	O
Dilleniaceae	*Tetracera scandens (L.) Merr.*		6658, 6668	fr	Pi	L	ind	F
Ebenaceae	*Diospyros ferrea (Willd.) Bakh.*	C	6813	st	CF	T	ind	O
Ebenaceae	*Diospyros lanceifolia Roxb.*		6701	st	MF	T	ind	O
Elaeocarpaceae	*Elaeocarpus floribundus Blume var. floribundus*		6727	st	MF	T	ind	F
Ericaceae	*Vaccinium leptanthum Miq.*		6726	st	Pi	E	ind	F
Erythroxylaceae	*Erythroxylum cuneatum (Miq.) Kurz*		6723	st	CF	T	ind	F
Euphorbiaceae	*Acalypha indica L.*		6846	fert	An	H	ind?	F
Euphorbiaceae	*Antidesma ghaesembilla Gaertn.*		6751	st	2F	T	ind	R
Euphorbiaceae	*Antidesma velutinosum Blume*		6625	st	MF	S	ind	O
Euphorbiaceae	*Austrobuxus nitidus Miq. var. nitidus*		6711	fr	CF	T	ind	O
Euphorbiaceae	*Baccaurea sp.1*		6700	st	MF	S	ind	F
Euphorbiaceae	*Breynia racemosa (Blume) Müll. Arg.*		6750, 6861	fr	Pi	S	ind	O
Euphorbiaceae	*Codiaeum variegatum L.*		o.r.		An	S	cult	F
Euphorbiaceae	*Endospermum diadenum (Miq.) Airy Shaw*		6802	st	MF	T	ind	O
Euphorbiaceae	*Euphorbia heterophylla L.*		6774	fr	An	H	nat	F
Euphorbiaceae	*Euphorbia hirta L.*		6770	fl, fr	An	H	nat	C
Euphorbiaceae	*Macaranga javanica (Blume) Müll. Arg.*		6737	st	An, Pi	T	ind?	F
Euphorbiaceae	*Macaranga tanarius (L.) Müll. Arg.*		6809	st	CF	T	ind?	O
Euphorbiaceae	*Manihot esculenta Crantz*		o.r.		An	S	cult	F
Euphorbiaceae	*Margaritaria indica (Dalzell) Airy Shaw*		6661	fl	CF	T	ind	F
Euphorbiaceae	*Phyllanthus amarus Schum. & Thonn.*		6780	fl, fr	An	H	nat	C
Euphorbiaceae	*Phyllanthus oxyphyllus Miq.*		6746, 6870	fl, fr	MF	S	ind	O
Euphorbiaceae	*Sauropus androgynus (L.) Merr.*		6784	fl, fr	An	S	cult	R
Euphorbiaceae	*Suregada multiflora (Juss.) Baill. var. multiflora*	C	6640, 6749	fr	Co, 2F	T	ind	C
Fabaceae	*Alysicarpus vaginalis (L.) DC.*		6852	st	An	H	ind?	R
Fabaceae	*Archidendron bubalinum (Jack) I.C. Nielsen*		6603, 6623, 6730	fr	VF, MF	T	ind	F

Dicotyledons (Palaeodicots and Eudicots)

Name	Family Species	PR	Specimen n°	Compo	Hab.-Type	TB	Orig	Rarity
Fabaceae	*Baphia nitida Lodd.*		6786	fl	An	S	cult	C
Fabaceae	*Bauhinia kockiana Korth.*		o.r.		An	L	cult	F
Fabaceae	*Bauhinia semibifida Roxb. var. semibifida*		6716	st	VF	L	ind	O
Fabaceae	*Caesalpinia bonduc (L.) Roxb.*	CQ	6804	st	Co	S	ind	O
Fabaceae	*Callerya atropurpurea (Wall.) Schot*		6790	st	An	T	cult	R
Fabaceae	*Canavalia rosea (Sw.) DC.*	Q	6806	fl, fr	Co	V	ind	O
Fabaceae	*Dalbergia candenatensis (Dennst.) Prain*		6810	fl	Co	S	ind	R
Fabaceae	*Delonix regia (Hook.f.) Raf.*		o.r.		An	T	cult	O
Fabaceae	*Dendrolobium umbellatum (L.) Benth.*	C	6649	fl, fr	Co	S	ind	C
Fabaceae	*Derris trifoliata Lour.*		6823	st	Co	L	ind	F
Fabaceae	*Desmodium triflorum (L.) DC.*		6856	st	An	H	nat?	C
Fabaceae	*Entada spiralis Ridl.*		6764	st	MF	L	ind	R
Fabaceae	*Erythrina variegata L.*	C			CF	T	ind	
Fabaceae	*Leucaena leucocephala (Lam) De Wit*	Q	6777	fr	An	T	nat	C
Fabaceae	*Mimosa pudica L.*		o.r.		An	H	nat	F
Fabaceae	*Pongamia pinnata (L.) Pierre var. pinnata*		6824	fl, fr	CF	T	ind	O
Fabaceae	*Tamarindus indica L.*		6778	st	An	T	cult	O
Fabaceae	*Vigna marina (Burm.) Merr.*		6808	fl, fr	Co	V	ind	O
Fabaceae	*Fabaceae sp.1*		6630	st	MF	L	ind	F
Goodeniaceae	*Scaevola sericea Vahl*	CQ	6650	fl, fr	Co	S	ind	C
Hernandiaceae	*Hernandia nymphaeifolia (Presl) Kubitzki*		o.r.		CF	T	ind	R
Ixonanthaceae	*Ixonanthes reticulata Jack*		6597	fr	DF	T	ind	O
Lamiaceae	*Ocimum basilicum L.*		6895	fr	An	H	cult	R
Lamiaceae	*Ocimum tenuiflorum L.*		6892	fl, fr	An	H	nat	F
Lauraceae	*Cassytha filiformis L.*	CQ	6651	flb, fr	Co	V	ind	C
Lauraceae	*Cinnamomum iners Reinw. ex Bl. var. subcuneatum*		6897	st	An	T	cult	F
Lauraceae	*Litsea elliptica Blume*		6797	st	VF	T	ind	R
Lauraceae	*Litsea johorensis Gamble*		6739	st	MF	T	ind	R
Lauraceae	*Neolitsea zeylanica (Nees) Merr.*		6762	st	CF	T	ind	F
Lecythidaceae	*Barringtonia asiatica (L.) Kurz*	CQ	6828	st	CF	T	ind	F
Lecythidaceae	*Barringtonia macrostachya (Jack) Kurz*	C	6639	st	MF	T	ind	C
Loganiaceae	*Fagraea auriculata Jack subsp. auriculata*		6814	st	CF	T	ind	R
Lythraceae	*Pemphis acidula Forst.*	C	6601	fl, fr	Ma	T	ind	C
Lythraceae	*Sonneratia griffithii Kurz*		6880	fr	Ma	T	ind	R
Malpighiaceae	*Tristellateia australasiae A. Rich.*	C			Ma	V	ind?	
Malvaceae	*Durio zibethinus Rumph. ex Murray*		o.r.		An	T	cult	O
Malvaceae	*Heritiera javanica (Blume) Kosterm.*		6709	st	MF	T	ind	O
Malvaceae	*Heritiera littoralis Dryand.*	C			CF	T	ind	
Malvaceae	*Hibiscus mutabilis L*		o.r.		An	S	cult	R
Malvaceae	*Hibiscus tiliaceus L.*	CQ	6827	fl, fr	CF	T	ind	F
Malvaceae	*Hibiscus tiliaceus L. 'Hilo Rainbow'*		o.r.		An	S	cult	O
Malvaceae	*Scaphium macropodum (Miq.) Beumée ex Heyne*		6867	st	VF	T	ind	R
Melastomataceae	*Clidemia hirta (L.) D. Don*		6868	st	VF	S	inv	R
Melastomataceae	*Melastoma malabathricum L.*	Q	6713	fl, fr	Pi	S	ind?	C
Melastomataceae	*Memecylon ovatum Sm.*		6683	fr	CF	T	ind	F
Meliaceae	*Azadirachta indica A. Juss.*		6843	st	An	T	cult	F
Meliaceae	*Lansium domesticum Corrêa*		6889	st	An	T	cult	R
Menispermaceae	*Fibraurea tinctoria Lour.*		6644	st	MF	L	ind	C
Menispermaceae	*Pericampylus glaucus (Lam.) Merr.*		6653	st	Co	V	ind?	O

Dicotyledons (Palaeodicots and Eudicots)

Name	Family Species	PR	Specimen n°	Compo	Hab.-Type	TB	Orig	Rarity
Moraceae	*Artocarpus dadah Miq.*		6755, 6885	fr	MF	T	ind	O
Moraceae	*Artocarpus heterophyllus Lam.*		o.r.		An	T	cult	O
Moraceae	*Artocarpus integer (Thunb.) Merr. var. integer*		o.r.	fr	An	T	cult	R
Moraceae	*Ficus caulocarpa Miq.*		6652	fr	CF	T	ind	C
Moraceae	*Ficus fistulosa Reinw. ex Blume*		6796	st	VF, An	T	ind	F
Moraceae	*Ficus grossularioides Burm.f.*		6675	fr	MF	S	ind	F
Moraceae	*Ficus vasculosa Wall. ex Miq.*		6708	st	MF	S	ind	O
Myristicaceae	*Knema globularia (Lam.) Warb.*		6620, 6763	flb, fr	MF, VF	T	ind	F
Myrsinaceae	*Ardisia elliptica Thunb.*		6634	flb, fr	Pi	S	ind?	F
Myrsinaceae	*Ardisia korthalsiana Scheff.*		6765	fl	MF	S	ind	R
Myrsinaceae	*Ardisia ridleyi King & Gamble*		6792	fl, fr	MF	S	ind	O
Myrtaceae	*Psidium guajava L.*		o.r.		An	T	cult	O
Myrtaceae	*Rhodamnia cinerea Jack*		6706	st	DF, MF	T	ind	C
Myrtaceae	*Syzygium aqueum (Burm.f.) Alston*		6842, 6866	fl	VF, An	T	nat	O
Myrtaceae	*Syzygium campanulatum Korth.*		6890	st	An	T	cult	F
Myrtaceae	*Syzygium grande (Wight) Walp.*	C	6628, 6629, 6631, 6635	flb, fr	MF, 2f	T	ind	A
Myrtaceae	*Syzygium zeylanicum (L.) DC.*		6860	fr	DF	T	ind	F
Nyctaginaceae	*Bougainvillea glabra Choisy*		o.r.		An	L	cult	F
Oleaceae	*Chionanthus ramiflorus Roxb.*		6622, 6710, 6791	fl, flb, fr	DF, CF	S	ind	F
Oleaceae	*Jasminum sambac (L.) Ait.*		6891	fl	An	S	cult	O
Opiliaceae	*Champereia manillana (Blume) Merr.*		6618, 6627, 6696, 6719	fl, fr	CF, MF	S	ind	F
Oxalidaceae	*Averrhoa carambola L.*		o.r.		An	T	cult	O
Pandaceae	*Microdesmis caseariifolia Planch.*		6617, 6637	st	MF	T	ind	F
Passifloraceae	*Passiflora foetida L.*		6811	fr	Co	V	nat	F
Piperaceae	*Peperomia pellucida (L.) H.B. & K.*		6902	fert	An	H	nat	O
Piperaceae	*Piper caninum Blume*		6681	st	VF	V	ind?	R
Polygalaceae	*Xanthophyllum vitellinum (Blume) Dietr.*		6733	st	MF	T	ind	R
Rhamnaceae	*Colubrina asiatica (L.) Brongn.*		6646	fl, fr	Co	S	ind	C
Rhizophoraceae	*Bruguiera gymnorhiza Lam.*	Q	6883	fl, fr	Ma	T	ind	R
Rhizophoraceae	*Rhizophora mucronata Lam.*	Q	6600	flb	Ma	T	ind	C
Rubiaceae	*Canthium confertum Korth.*		6734	st	MF	S	ind	O
Rubiaceae	*Chassalia chartacea Craib*		6595, 6621, 6695	fr	MF	S	ind	F
Rubiaceae	*Diplospora malaccensis Hook.f.*		6697	st	MF	S	ind	R
Rubiaceae	*Guettarda speciosa L.*	CQ	6662	st	CF	T	ind	F
Rubiaceae	*Gynochthodes sublanceolata Miq.*		6712	fr	Pi	V	ind	F
Rubiaceae	*Hedyotis capitellata Wall. ex G. Don*		6736	fr	Pi	V	ind	F
Rubiaceae	*Hedyotis philippinensis (Willd. ex Spreng.) Merr. ex C.B. Rob.*		6864	fert	VF	S	ind	R
Rubiaceae	*Ixora coccinea L.*		o.r.		An	S	cult	F
Rubiaceae	*Ixora congesta Roxb.*		6604	flb	MF, VF	S	ind	F
Rubiaceae	*Ixora lobbii King & Gamble*		6615	st	MF	S	ind	C
Rubiaceae	*Ixora pendula Jack var. pendula*		6705	fr	MF	S	ind	O
Rubiaceae	*Lasianthus cynocarpus Jack*		6636, 6677	fl, fr	MF	S	ind	C
Rubiaceae	*Lasianthus pergamaceus King & Gamble*		6871	st	MF	T	ind	F
Rubiaceae	*Morinda citrifolia L.*	CQ	6647	fl, fr	CF	T	ind?	F
Rubiaceae	*Oldenlandia corymbosa L.*		6769, 6854	fl, fr	An	H	ind?	C

Dicotyledons (Palaeodicots and Eudicots)

Name	Family Species	PR	Specimen n°	Compo	Hab.-Type	TB	Orig	Rarity
Rubiaceae	*Paederia verticillata Blume*		6645	st	MF	L	ind	R
Rubiaceae	*Pavetta multiflora (Koord. & Valeton) Bremek.*		6607	fr	MF	S	ind	O
Rubiaceae	*Prismatomeris glabra (Korth.) Valeton*		6793	st	DF	S	ind	R
Rubiaceae	*Psychotria obovata Wall.*		6633	fl, fr	MF, CF	H	ind	C
Rubiaceae	*Spermacoce assurgens Ruiz & Pav.*		6835	fl, fr	An	H	nat	C
Rubiaceae	*Timonius compressicaulis (Miq.) Boerl.*	C	6656	fr	CF	T	ind	F
Rubiaceae	*Timonius wallichianus (Korth.) Valeton*		6803	st	MF	T	ind	O
Rutaceae	*Atalantia monophylla (L.) DC.*	CQ	6660	st	CF	T	ind	C
Rutaceae	*Maclurodendron porteri (Hook.f) T.G. Hartley*		6725	st	MF	T	ind	O
Rutaceae	*Murraya koenigii (L.) Spreng.*		6783	st	An	S	cult	O
Rutaceae	*Murraya paniculata (L.) Jack*		6855	fr	An	S	cult	F
Sapindaceae	*Allophyllus cobbe (L.) Raeusch.*	C			Co	S	ind	
Sapindaceae	*Dimocarpus longan Lour. subsp. malesianus Leenh.*		6894	st	An	T	cult	F
Sapindaceae	*Dodonaea viscosa Jacq.*		6773	st	An	S	ind	R
Sapindaceae	*Guioa pleuropteris (Blume) Radlk.*		6669, 6731	flb, fr	CF, Pi	T	ind	F
Sapindaceae	*Lepisanthes fruticosa (Roxb.) Leenh.*		6704, 6747	st	MF	T	ind	F
Sapindaceae	*Nephelium lappaceum L.*		6888	st	An	T	cult	F
Sapotaceae	*Manilkara zapota (L.) P. Royen*		o.r.		An	T	cult	F
Sapotaceae	*Palaquium obovatum (Griff.) Engl.*		6608, 6863, 6900	fl	MF	T	ind	C
Sapotaceae	*Planchonella obovata (R. Br.) Pierre*	C	6812, 6881	st	CF	T	ind	F
Scrophulariaceae	*Lindernia crustacea (L.) F. Muell.*		6851	fl, fr	An	H	ind	O
Simaroubaceae	*Eurycoma longifolia Jack*		6715	st	MF	T	ind	F
Solanaceae	*Physalis minima L.*		6884	fl, fr	An	H	ind	O
Ulmaceae	*Gironniera subaequalis Planch.*		6611	st	MF	T	ind	F
Urticaceae	*Pilea microphylla (L.) Liebm.*		o.r.		An	H	nat	F
Verbenaceae	*Clerodendrum myrmecophila Ridl.*		6759	st	MF	S	ind	O
Verbenaceae	*Lantana camara L.*	Q	6768	fl	An, Pi	S	inv	C
Verbenaceae	*Premna serratifolia L.*	C	6599	fl, fr	Co	T	ind	C
Verbenaceae	*Teijsmanniodendron coriaceum (C.B. Clarke) Kosterm.*		6799	st	VF	T	ind	R
Verbenaceae	*Vitex pinnata L.*	C	6663	st	CF	T	ind	C
Verbenaceae	*Vitex trifolia L. var. repens Ridley*		6820	fr	Co	S	ind	O
Vitaceae	*Cissus hastata Miq.*		6752	st	Pi	V	ind	O
Vitaceae	*Leea indica (Burm.f.) Merr.*		6692	fr	VF, Pi	T	ind	O

Monocotyledons

Name	Family Species	PR	Specimen n°	Compo	Hab.-Type	TB	Orig	Rarity
Agavaceae	*Cordyline fruticosa (L) A. Chev.*		o.r.		An	S	cult	O
Amaryllidaceae	*Amaryllidaceae sp.1*		o.r.		An	H	cult	F
Araceae	*Alocasia longiloba Miq.*		o.r.		An	H	cult	O
Araceae	*Alocasia macrorrhizos (L.) G. Don*		o.r.		An	H	nat	F
Araceae	*Dieffenbachia seguine (Jacq.) Schott*		o.r.		An	H	nat	O
Araceae	*Epipremnum pinnatum (L.) Engl.*		o.r.		An	V	nat	O
Araceae	*Homalomena sagittifolia Jungh. ex Schott*		6865, 6887	fl	VF	H	ind?	O
Araceae	*Philodendron bipinnatifidum Schott ex Endl.*		o.r.		An	H	cult	O
Arecaceae	*Areca catechu L.*		6841	st	An	T	cult	F
Arecaceae	*Arenga brevipes Becc.*		6679	st	VF	H	ind	O
Arecaceae	*Calamus burkillianus Becc. ex Ridl.*	Q	6670	fl, fr	MF, VF	L	end	C
Arecaceae	*Caryota mitis Lour.*	Q	6682	st	VF, An	T	ind	F
Arecaceae	*Chrysalidocarpus lutescens Wendl.*		6839	st	An	T	cult	F
Arecaceae	*Cocos nucifera L.*	CQ	o.r.		CF, An	T	nat	F
Arecaceae	*Licuala grandis H. Wendl.*		o.r.		An	S	cult	F
Arecaceae	*Licuala spinosa Wurmb.*		6858	st	MF	H	ind	R
Arecaceae	*Metroxylon sagu Rottb.*	Q	6690	fr	VF	T	cult	O
Arecaceae	*Oncosperma tigillarium (Jack) Ridl.*	Q	6678	st	MF, VF	T	ind	F
Arecaceae	*Ptychosperma macarthurii (Wendl.) Nichols.*		6840	fr	An	T	cult	F
Arecaceae	*Rhapis excelsa (Thunb.) Henry*		6833	st	An	S	cult	F
Arecaceae	*Rhapis multifida Burret*		o.r.		An	S	cult	O
Commelinaceae	*Tradescantia spathacea Swartz*		o.r.		An	H	cult	R
Cyperaceae	*Cyperus radians Nees & Meyen*		6876	fert	Co	H	ind	O
Cyperaceae	*Cyperus rotundus L.*		6844, 6850	fert	An	H	nat?	C
Cyperaceae	*Fimbristylis cymosa R. Br.*		6822	fert	Co	H	ind	F
Cyperaceae	*Kyllinga nemoralis (J.R. Forst. & G. Forst.) Dandy ex Hutch. & Dalziel*		6893	fert	An	H	nat?	F
Cyperaceae	*Mapania palustris (Hassk. ex Steud.) Fern.-Vill. var. palustris*		6800	st	VF	H	ind	R
Cyperaceae	*Remirea maritima Aubl.*		6878	st	Co	H	ind	O
Cyperaceae	*Scleria lithosperma (L.) Sw.*		6673, 6869	fr	MF, CF, DF	H	ind	F
Cyperaceae	*Scleria oblata S.T. Blake*		6602, 6862	fert	VF, MF	H	ind?	F
Cyperaceae	*Scleria sumatrensis Retz.*		6688	fl, fr	CS	H	ind	F
Dioscoreaceae	*Tacca palmata Blume*		6606, 6667	fl, fr	MF, CF	H	ind	C
Dioscoreaceae	*Tacca pinnatifida J.R. Forst. & G. Forst.*		6787	fr	CF	H	ind?	R
Flagellariaceae	*Flagellaria indica L.*		6698	st	VF	L	ind	F
Heliconiaceae	*Heliconia psittacorum Lf.*		o.r.		An	H	cult	R
Heliconiaceae	*Heliconia rostrata Ruiz & Pav.*		o.r.		An	H	cult	F
Hemerocallidacea e	*Dianella ensifolia (L.) DC.*	C	6598, 6718	fl, fr	DF	H	ind	F
Iridaceae	*Trimezia martinicensis (Jacq.) Herb.*		o.r.		An	H	cult	R
Marantaceae	*Donax grandis (Miq.) K. Schum.*		6886	st	An	H	ind?	O
Musaceae	*Musa spp.*		o.r.		An	H	cult	F
Orchidaceae	*Aerides odorata Lour.*		6818	st	CR	E	ind	R
Orchidaceae	*Arachnis hookeriana (Reichb.f.) Rolfe*		6826	fl	CR	H	ind	O
Orchidaceae	*Cymbidium finlaysonianum Lindl.*	C	6815	st	CR	H	ind	O
Orchidaceae	*Dendrobium crumenatum Sw.*	C	6817	fl	CR	E	ind	O
Orchidaceae	*Orchidaceae sp.1*		6816	st	CR	E	ind	O
Pandanaceae	*Pandanus amaryllifolius Roxb.*		6781	st	An	S	cult	F

Monocotyledons

Name	Family Species	PR	Specimen n°	Compo	Hab.-Type	TB	Orig	Rarity
Pandanaceae	*Pandanus odoratissimus L.f.*	CQ	6760	st	DF, CF	T	ind	F
Pandanaceae	*Pandanus sandens Holt.*		o.r.		An	T	cult	R
Pandanaceae	*Pandanus tetrodon Ridl.*		6703	st	MF	S	ind	F
Poaceae	*Bambusa sp.1*		6834	st	An	S	cult	C
Poaceae	*Centotheca lappacea (L.) Desv.*		6691	fl	VF	H	ind	O
Poaceae	*Chrysopogon zizanioides (Linnaeus) Roberty*		6898	fert	An	H	cult	F
Poaceae	*Cymbopogon citratus (DC.) Stapf*		o.r.		An	H	cult	O
Poaceae	*Dactyloctenium aegyptium (L.) Willd.*		6779	fert	An	H	ind	C
Poaceae	*Digitaria longiflora (Retz.) Pers.*		6772	fert	An	H	ind?	F
Poaceae	*Eleusine indica (L.) Gaertn.*		6771	fl, fr	An	H	ind?	C
Poaceae	*Eragrostis amabilis (L.) Hook. & Arn.*		6788	fl	An	H	nat	F
Poaceae	*Eulalia ridleyi (Hack.) A. Camus*	C	6882	st	Co	H	end	O
Poaceae	*Imperata cylindrica (L.) Beauv.*	Q	6687	fl	CS	H	ind?	C
Poaceae	*Ischaemum muticum L.*	C	6684, 6874, 6879	fl	CS, Co	H	ind	F
Poaceae	*Lepturus repens (G. Forst) R. Br.*		6877	fert	Co	H	ind	O
Poaceae	*Paspalum vaginatum Sw.*		6875	fl	Co	H	ind	O
Poaceae	*Saccharum officinarum L.*		o.r.		An		cult	O
Poaceae	*Schizachyrium brevifolium (Sw.) Nees ex Büse*	C			Co	H	ind	
Poaceae	*Spinifex littoreus (Burm.f.) Merr.*	C			Co	H	ind	
Poaceae	*Zoysia matrella (L.) Merr.*		6853	fl	An	H	ind	F
Ruscaceae	*Dracaena elliptica Thunb.*		6613	st	MF	S	ind	F
Ruscaceae	*Dracaena maingayi Hook.f.*		6707	st	MF, CF	S	end	F
Stemonaceae	*Stemona curtisii Hook.f.*		6728	st	VF	V	ind	F
Zingiberaceae	*Etlingera elatior (Jack) R.M. Sm.*		o.r.		An	H	cult	R

Gymnosperms

Name	Family Species	PR	Specimen n°	Compo	Hab.-Type	TB	Orig	Rarity
Cycadaceae	*Cycas edentata de Laub.*	CQ	6821	fr	CR	T	ind	F
Gnetaceae	*Gnetum gnemon L. var. gnemon*		6643, 6724	fl	MF	T	ind	C
Gnetaceae	*Gnetum latifolium Blume var. funiculare (Blume) Markgr.*		6766	st	MF	L	ind	O
Podocarpaceae	*Podocarpus polystachyus R. Br. ex Endl.*	Q	6596	fr	DF, CR	T	ind	C

Ferns and Lycophytes

Name	Family Species	PR	Specimen n°	Compo	Hab.-Type	TB	Orig	Rarity
Aspleniaceae	*Asplenium nidus L.*		6741	fert	CF, VF, An	E	ind	F
Blechnaceae	*Blechnum finlaysonianum Hook. & Grev.*		6729	fert	An, Pi	H	ind	F
Blechnaceae	*Stenochlaena palustris (Burm.) Bedd.*		6789	st	CS, VF	V	ind	F
Davalliaceae	*Davallia denticulata (Burm.f.) Mett. ex Kuhn var. denticulata*		6740	fert	DF, CR	E	ind	F
Dennstaedtiaceae	*Pteridium caudatum (L.) Maxon subsp. yarrabense (Domin) Parris*		6674	fert	Pi	H	ind	R
Gleicheniaceae	*Dicranopteris linearis (Burm.f.) Underw.*	Q	6830	fert	Pi	H	ind	F
Lindsaeaceae	*Lindsaea ensifolia Sw.*		6671	fert	VF	H	ind	F
Lycopodiaceae	*Lycopodiella cernua (L.) Pic. Serm.*	Q	6798	fert	CS, Pi	H	ind	O
Lygodiaceae	*Lygodium flexuosum (L.) Sw.*		6659, 6753	fert	An, Pi	V	ind	F
Lygodiaceae	*Lygodium microphyllum (Cav.) R. Br.*		6689	fert	Pi, CS	V	ind	F
Nephrolepidaceae	*Nephrolepis acutifolia (Desv.) Christ*		6721	fert	CR	E	ind	R
Nephrolepidaceae	*Nephrolepis biserrata (Sw.) Schott*		6657	fert	Co, An	H	ind	F
Nephrolepidaceae	*Nephrolepis cordifolia (L.) C.Presl var. cordifolia*		6754	st	An	H	nat	R
Polypodiaceae	*Drynaria quercifolia (L.) J.Sm.*		6743	st	DF, CR	E	ind	R
Polypodiaceae	*Phymatodes scolopendria (Burm.) Ching*		6742	st	CF, CR	H	ind	O
Polypodiaceae	*Platycerium bifurcatum (Cav.) C. Chr.*		o.r.		An	E	cult	F
Polypodiaceae	*Pyrrosia lanceolata (L.) Farw.*		6720	fert	CR, Pi	E	ind	F
Polypodiaceae	*Pyrrosia piloselloides (L.) M.G. Price*		6676, 6738	fert	MF	E	ind	O
Pteridaceae	*Acrostichum aureum L.*		o.r.		Co	H	ind	R
Pteridaceae	*Adiantum latifolium Lam.*		6665	fert	CF, CR	H	nat	R
Pteridaceae	*Adiantum stenochlamys Baker*		6722	fert	CF, CR	H	ind	O
Pteridaceae	*Pityrogramma calomelanos (Linn.) Link*		6782	fert	An	H	nat	O
Pteridaceae	*Pteris ensiformis Burm. f.*		6664	fert	An	H	nat	R
Pteridaceae	*Pteris tripartita Sw.*		6901	fert	An	H	ind?	R
Pteridaceae	*Pteris vittata Linn.*		6832	fert	An	H	nat	R
Pteridaceae	*Taenitis blechnoides (Willd.) Sw.*		6693	fert	VF	H	ind	F
Schizaeaceae	*Actinostachys digitata (L.) Wall. ex Reed*		6672	fert	DF, CR	H	ind	R
Schizaeaceae	*Schizaea dichotoma (L.) Sm.*		6819	fert	CR	H	ind	R
Thelypteridaceae	*Cyclosorus dentatus (Forssk.) Ching*		6666	fert	An, Pi	H	ind	F
Thelypteridaceae	*Pronephrium asperum (Presl) Holtt.*		6680, 6694	fert	VF	H	ind	O

List of the Terrestial Vascular Plants and their uses and local names.

The most interesting species are emphasized in bold.

Species Name	Venacular Names (Malay)	Uses
Acalypha indica	L.Rumput lislis	
Acrostichum aureum L.	Piai raya, Paku laut, Larat	
Actinostachys digitata (L.) Wall. ex Reed		Aphrodisiac and to relieve coughing
Adiantum latifolium Lam.		
Adiantum stenochlamys Baker		
Aerides odorata Lour.		Ornamental
Ageratum conyzoides L.	Tahi ayam, Tombok jantan, Sianggit	
Allamanda cathartica L.		
Allophyllus cobbe (L.) Raeusch.		
Alocasia longiloba Miq.		
Alocasia macrorrhizos (L.) G. Don		
Alstonia spatulata Blume	Pulai paya	Very light wood
Alternanthera sessilis (L.) R.Br. ex Roth	Daun tolod	Intestinal cramps
Alysicarpus vaginalis (L.) DC.		
Alyxia reinwardtii Blume		Fragrant: the whole plant smells of coumarin and the dried bark was formerly put among clothes to perfume them
Amaranthus dubius Mart. ex Thell.	Bayam	
Amaryllidaceae sp.1		
Anacardium occidentale L.	Gajus, Jambu golok, Keterek	
Antidesma ghaesembilla Gaertn.	Gunchiak, Balong ayam	
Antidesma velutinosum Blume	Mempunai bukit, Sutupo bukit, Burubah rimbah	
Arachnis hookeriana (Reichb.f.) Rolfe	White scorpion orchid	Ornamental
Archidendron bubalinum (Jack) I.C. Nielsen	Giring antan, Gerdas padi, Keredas	Seeds used for flavouring food; timber said to be durable (FMI11: 97)
Ardisia elliptica Thunb.	Lutus	
Ardisia korthalsiana Scheff.		
Ardisia ridleyi King & Gamble		
Areca catechu L. Pinang	Betelnut	
Arenga brevipes Becc.		
Arivela viscosa (L.) Raf. var. viscosa	Kutepeng, maman pantai	Relieves headaches, also for colic and dysentery
Arthrophyllum diversifolium Blume	Tumboh kelapa	
Artocarpus dadah Miq.	Tampang bulu	
Artocarpus heterophyllus Lam.	Nangka	
Artocarpus integer (Thunb.) Merr. var. integer	Cempedak	
Asplenium nidus L.		
Asystasia gangetica (L.) T. Anderson subsp. micrantha (Nees) Ensermu		
Atalantia monophylla (L.) DC.	Akar merlimau, Empenai	
Austrobuxus nitidus Miq. var. nitidus		
Averrhoa carambola L.	Belimbing	
Azadirachta indica A. Juss.	Ranggu	Contains Azadirachtin which is an insecticide. The living tree is an effective deterrent against mosquitoes.
Baccaurea sp.1		
Bambusa sp.1		
Baphia nitida Lodd.		
Barringtonia asiatica (L.) Kurz	Putat laut, Butun, Pertun, Butung	
Barringtonia macrostachya (Jack) Kurz	Putat gajah, Putat hutan	

Species Name	Venacular Names (Malay)	Uses
Bauhinia kockiana Korth.		
Bauhinia semibifida Roxb. var. semibifida		
Blechnum finlaysonianum Hook. & Grev.		
Blumea balsamifera (L.) DC.	Chapa, Chapu, Sembong	Ngai camphor
Bougainvillea glabra Choisy		
Breynia racemosa (Blume) Müll. Arg.	Poko ambin kra, Sumbar, Saga, Hujan panas	
Bruguiera gymnorhiza Lam.	Lenggadai, Tumu merah	
Buchanania arborescens (Blume) Blume	Katak hudang, Kelumpang keras, etc.	
Caesalpinia bonduc (L.) Roxb.	Olang, K'lubu	Seeds are used as decorative objects.
Calamus burkillianus Becc. ex Ridl.	Ratan	
Callerya atropurpurea (Wall.) Schot	Tulang daeng, Chicha, Girah payah, Merbong	
Calophyllum depressinervosum M.R. Hend. & Wyatt-Sm.	Bintangor lekok	
Calophyllum inophyllum L.	Bintanggor laut	
Canarium pilosum Benn.	Kadondon krut, Kadondon hutan	
Canavalia rosea (Sw.) DC.		Seeds edible when young. They makes good pea porridge. The flowers eaten raw by Malays.
Canthium confertum Korth.		
Capparis micracantha DC. subsp. korthalsiana (Miq.) M. Jacobs	Jambol merah	
Carica papaya L.	Betik	
Caryota mitis Lour.	Tukus, Dudok	The stems produce an inferior sago. Ornamental.
Cassytha filiformis L.		
Casuarina equisetifolia L.	Ru, Aru, Eru	
Catharanthus roseus (L.) G. Don	Kemunting cina	
Centotheca lappacea (L.) Desv.	Rumpur darah, Rumput silat kain	
Cerbera manghas L.	Pokok pong pong	Seeds can be made into a fish poison.
Champereia manillana (Blume) Merr.	Buah chemperai Chassalia chartacea Craib	
Chionanthus ramiflorus Roxb.		
Chromolaena odorata (L.) R.M. King. & H. Robuns.		
Chrysalidocarpus lutescens Wendl.		
Chrysopogon zizanioides (Linnaeus) Roberty	Vetiver	Cultivated for its aromatic roots.
Cinnamomum iners Reinw. ex Bl. var. subcuneatum (Miq.) W.K. Soh	Lelang, Kayu manis hutan	Commonly planted tree.
Cissus hastata Miq.		
Clerodendrum myrmecophila Ridl.		
Clidemia hirta (L.) D. Don		
Clusia rosea L.		
Cocos nucifera L.	Kelapa	
Codiaeum variegatum L.	Puding	
Colubrina asiatica (L.) Brongn.	Periah pantai	
Connarus monocarpus L. subsp. malayensis Leenh.	Akar tulang dayeng	
Cordia subcordata Lam.		
Cordyline fruticosa (L) A.Chev.		
Cycas edentata de Laub.	Paku laut	Ornamental; medicinal, alimentary.
Cyclosorus dentatus (Forssk.) Ching		
Cymbidium finlaysonianum Lindl.	Sakat panggan daun, Bunga chandarek	Ornamental
Cymbopogon citratus (DC.) Stapf		
Cynanchum ovalifolium Wight	Akar bano	
Cyperus radians Nees & Meyen		
Cyperus rotundus L.	Rumput haliya hitam	The tubers are used for smoking in cases of pain in the nose.
Dactyloctenium aegyptium (L.) Willd.		

Species name	Venacular Names (Malay)	Uses
Dalbergia candenatensis (Dennst.) Prain		
Davallia denticulata (Burm.f.) Mett. ex Kuhn var. denticulata	Paku tertutup	
Delonix regia (Hook.f.) Raf.	Poinciana	
Dendrobium crumenatum Sw.	Poko merpati, Bunda angin, Pigeon orchid	Ornamental
Dendrolobium umbellatum (L.) Benth.)	Dedulang	
Derris trifoliata Lour.	Akar ketuil	
Desmodium triflorum (L.) DC		
Desmos dasymaschalus (Blume) Safford var. wallichii (Hook.f. & Thomson) Ridl.	Jari ayam, Kenanga paya, Chenang hutan, Larut api, Chin ching	
Dianella ensifolia (L.) DC.	Satagit, Siak-siak jantan, Senjuang, Benjuan	Good potential as ornamental plant.
Dicranopteris linearis (Burm.f.) Underw.	Resam, Bengakawang	
Dieffenbachia seguine (Jacq.) Schott		
Digitaria longiflora (Retz.) Pers.		
Dillenia pulchella (Jack) Gilg	Simpur ayer	Wood is hard and heavy, suitable for construction.
Dimocarpus longan Lour. subsp. malesianus Leenh	Mata kuching	
Diospyros ferrea (Willd.) Bakh.	Kaya arang, Sachirit laut	
Diospyros lanceifolia Roxb.		
Diplospora malaccensis Hook.f.	Uloh-uloh	Leaves used in place of coffee leaves for coffee-leaf tea.
Dischidia major (Vahl) Merr.	Akar kul, Akar bano	Good potential as ornamental plant.
Dodonaea viscosa Jacq.	Kayu bertih, Seringan laut	
Donax grandis (Miq.) K. Schum.	Bemban	
Dracaena elliptica Thunb.		Good potential as ornamental plant.
Dracaena maingayi Hook.f.		Good potential as ornamental plant.
Drynaria quercifolia (L.) J. Sm.	Daun kepala tupai, Skat laipang	
Durio zibethinus Rumph. ex Murray	Durian	
Elaeocarpus floribundus Blume var. floribundus	Mendong, Medang, kelawar	Timber
Eleusine indica (L.) Gaertn.	Rumput sambau	
Emilia sonchifolia (L.) DC.	Katumbit jantan, Satumbak merah	
Endospermum diadenum (Miq.) Airy Shaw	Bebaru bukit, Sendok-sendok, Sesundo	The soft white wood is used for making clogs.
Entada spiralis Ridl.		
Epipremnum pinnatum (L.) Engl.		
Eragrostis amabilis (L.) Hook. & Arn.	Rumput tulong belalang	
Erythrina variegata L.		Often planted in urban areas.
Erythroxylum cuneatum (Miq.) Kurz	Cinta mula, Ketai mula, Baka, Inai inai	
Etlingera elatior (Jack) R.M. Sm.	Kantan	Largely cultivated for its flower shoots that are used in curry.
Eulalia ridleyi (Hack.) A. Camus		
Euphorbia heterophylla L.		
Euphorbia hirta L.	Ambin jantan, Kurumak susu, Gelang susu	
Eurycoma longifolia Jack	Tongkat ali	Medicinal, very widely used and harvested in the wild. Among its uses: blood coagulant for complications during childbirth, as a treatment for dysentery, and as an aphrodisiac.
Fabaceae sp.1		
Fagraea auriculata Jack subsp. auriculata	Temasuk, Tembusu?	Good potential as ornamental plant.
Fibraurea tinctoria Lour.	Akar kinching kerbau, Akar kuning	Medicinal
Ficus caulocarpa Miq.Ornamental		
Ficus fistulosa Reinw. ex Blume		
Ficus grossularioides Burm.f.	Chumantong, Kulupong burong, Kulupong ayer	
Ficus vasculosa Wall. ex Miq.	Ara Nasi, Supudi Tampang Burong	

Species name	Venacular Names (Malay)	Uses
Fimbristylis cymosa R. Br.		
Flagellaria indica L. Rotan bini	Rotan tikus	Basket-making.
Garcinia mangostana L		
Garcinia nigrolineata Planch. ex T. Anderson	Kandis jantan	
Garcinia parvifolia (Miq.) Miq.		
Gironniera subaequalis Planch.	Hampas tebu, Medang kasap, Medang bulanak	
Gnetum gnemon L. var. gnemon	Buah maninjau	The seeds are eaten like nuts.
Gnetum latifolium Blume var. funiculare (Blume) Markgr.	Akar tutubo, Akar suburus, Areuj kasungka, Tangkil	
Guettarda speciosa L.		
Guioa pleuropteris (Blume) Radlk.	Sempayan ulur, Pena-pena	
Gynochthodes sublanceolata Miq.	Akar lampai hitam	
Hedyotis capitellata Wall. ex G. Don	Kemeniyan hantu, Anga besi, Akar lidah jin, Sampu keladi, Keresek pisang, Sutnibut	
Hedyotis philippinensis (Willd. ex Spreng.) Merr. ex C.B. Rob.		
Heliconia psittacorum Lf. Heliconia rostrata Ruiz & Pav.		
Heritiera javanica (Blume) Kosterm.	Mengkulang jari	
Heritiera littoralis Dryand.	Dungun, Peler kambing, Atun laut, Bayur laut	Wood very hard and compact, used for firewood.
Hernandia nymphaeifolia (Presl) Kubitzki	Bua k'ras laut	
Hibiscus mutabilis L		
Hibiscus tiliaceus L.	Baru, Ambaru, Baru laut, Dedap laut	
Hibiscus tiliaceus L. 'Hilo Rainbow'		
Homalomena sagittifolia Jungh. ex Schott		
Hoya diversifolia Blume	Akar sesudu bukit, Akar chapang kra	Good potential as ornamental plant
Imperata cylindrica (L.) Beauv.	Lalang	The leaves are used for thatching and litter, the silky spikelets for stuffing pillows, the rhizomes are exported to China as a medicine, and formerly an ingredient for a Singaporean beer.
Ipomoea littoralis Blume		
Ipomoea pes-caprae (L.) Sweet	Tapah kudah	
Ischaemum muticum L.	Rumput tembaga jantan, Rumput terutus tembaga	Malays use the leaves for headaches.
Ixonanthes reticulata Jack	Jinjagong, Sakit hudang, Pagar anak	
Ixora coccinea L.		
Ixora congesta Roxb.	Jarum saluang, Bunga penaga riam	
Ixora lobbii King & Gamble	Kramat hujan, Pechah priok	
Ixora pendula Jack var. pendula	Jambol siol, Mumjilai hutan, Saratong padi, Tabong bunga	
Jasminum sambac (L.) Ait.		
Knema globularia (Lam.) Warb.	Chendarah padi	
Kyllinga nemoralis (J.R. Forst. & G. Forst.) Dandy ex Hutch. & Dalziel	Rumput butong	
Lansium domesticum Corrêa	Duku	
Lantana camara L.	Bunga tahi ayam	
Lasianthus cynocarpus Jack		
Lasianthus pergamaceus King & Gamble		
Leea indica (Burm.f.) Merr.		
Lepisanthes fruticosa (Roxb.) Leenh.		Medicinal
Lepturus repens (G.Forst) R. Br.		

Species name	Venacular Names (Malay)	Uses
Leucaena leucocephala (Lam) De Wit	Ipil-ipil (Philippines)	
Licuala grandis H. Wendl.		
Licuala spinosa Wurmb.		
Lindernia crustacea (L.) F. Muell.	Kerak nasi	
Lindsaea ensifolia Sw.		
Litsea elliptica Blume	Perawas	Highly valued in native medicine. A very aromatic tree.
Litsea johorensis Gamble		
Lycopodiella cernua (L.) Pic. Serm.	Paku serani, Rumput serani, Rumput kenarus	
medicinal, ornamental		
Lygodium flexuosum (L.) Sw.		
Lygodium microphyllum (Cav.) R. Br.		
Macaranga javanica (Blume) Müll. Arg.		
Macaranga tanarius (L.) Müll. Arg.	Kundo, Inchong	
Maclurodendron porteri (Hook.f) T.G. Hartley	Katiak, Bunan hutan, Melaman	
Mangifera indica L.	Mangga, Mempelam, Pauh	
Mangifera pentandra Hook.f.	Manga dodol, Mempelam bemban, Pauh, Pauh damar	Rather fragrant, watery flesh, pale orange, rather sweet with few fibers.
Manihot esculenta Crantz		
Manilkara zapota (L.) P. Royen		
Mapania palustris (Hassk. ex Steud.) Fern.-Vill. var. palustris	Harashas, Bangkonoh, Menkuang tedong, Menkuang lobo	The leaves are used for making mats and baskets.
Maranthes corymbosa Blume	Merbatu laut	Timber for house-building.
Margaritaria indica (Dalzell) Airy Shaw		
Melastoma malabathricum L.	Sendudok	Fruit used for thrush in children. Branches as an astringent in diarrhoea, roots as a mouth-wash for toothache.
Melothria pendula L.		
Memecylon ovatum Sm.		Good potential as ornamental plant
Metroxylon sagu Rottb.	Sago	Alimentation
Microdesmis caseariifolia Planch.	Chateng, Chereh rambeh, Kenidei badak, Sigoniah	
Mikania cordata (Burm.f.) Robinson	Akar lupang, Churoma, Akar ulam tikus	
Mimosa pudica L.		
Morinda citrifolia L.	Mengkudu	The fruit is eaten, though it has an unpleasant taste. Roots used for dyeing orange colour.
Murraya koenigii (L.) Spreng.		
Murraya paniculata (L.) Jack	Kamuning	
Musa spp.		
Neolitsea zeylanica (Nees) Merr.		
Nephelium lappaceum L.	Rambutan	
Nephrolepis acutifolia (Desv.) Christ		
Nephrolepis biserrata (Sw.) Schott	Paku uban, Paku larut	
Nephrolepis cordifolia (L.) C.Presl var. cordifolia		
Nerium oleander L.		
Ocimum basilicum L.	Ruku, Selasih antan	
Ocimum tenuiflorum L.		
Oldenlandia corymbosa L.	Telor B'lankas, Siku-siku, Sibu	For poulticing sores.
Oncosperma tigillarium (Jack) Ridl.	Nibong	Wood for building, pig spears, etc.
Orchidaceae sp.1		Ornamental
Paederia verticillata Blume		
Palaquium obovatum (Griff.) Engl.	Niato bunga, Niato tembaga, Niato balam, Taban balau putih, Taban balam kuning	

Species name	Venacular Names (Malay)	Uses
Pandanus amaryllifolius Roxb.		
Pandanus odoratissimus L.f.	Pandan duri, Pandan todas	The leaves are used for making mats and ataps.
Pandanus sandens Holt.		
Pandanus tetrodon Ridl.		
Parsonsia alboflavescens (Dennst.) Mabb.		
Paspalum vaginatum Sw.		
Passiflora foetida L.	Lang buluh, K'rang kraut, Letop-letop, Timun dindang, Timun padang	
Pavetta multiflora (Koord. & Valeton) Bremek.		
Pemphis acidula Forst.		
Peperomia pellucida (L.) H.B. & K.		
Pericampylus glaucus (Lam.) Merr.	Akar gasing-gasing, Akar gegasing, Akar jerkasing, Kelesu	
Philodendron bipinnatifidum Schott ex Endl.		
Phyllanthus amarus Schum. & Thonn.		Medicinal
Phyllanthus oxyphyllus Miq.	Cherek hantu	
Phymatodes scolopendria (Burm.) Ching	Paku wangi	Fragrant, was used to scent clothes and coconut oil.
Physalis minima L.	Chipluan	
Pilea microphylla (L.) Liebm.		
Piper caninum Blume	Chabai hutan, Lada hantu, Lada anjing, Akar kalong	
Pityrogramma calomelanos (Linn.) Link	Paku hijau	Medicinal, ornamental
Planchonella obovata (R. Br.) Pierre	Tuak-tuak	
Platycerium bifurcatum (Cav.) C. Chr.	Paku uncal	Ornamental, cultivated all over the world.
Plumeria obtusa L.		
Podocarpus polystachyus R. Br. ex Endl.	Podo laut	Ornamental
Pongamia pinnata (L.) Pierre var. pinnata	Malapari	
Premna serratifolia L.		
Prismatomeris glabra (Korth.) Valeton		
Pronephrium asperum (Presl) Holtt.		
Psidium guajava L.	Jambu biji	
Psychotria obovata Wall.		
Pteridium caudatum (L.) Maxon subsp. yarrabense (Domin) Parris		
Pteris ensiformis Burm.f.		
Pteris tripartita Sw.		
Pteris vittata Linn.	Paku uban bukit	
Ptychosperma macarthurii (Wendl.) Nichols.		
Pyrrosia lanceolata (L.) Farw.		
Pyrrosia piloselloides (L.) M.G. Price	Sisek naga	
Remirea maritima Aubl.		
Rhapis excelsa (Thunb.)Henry Rhapis multifida Burret		
Rhizophora mucronata Lam.	Belukap, Bakau kurap	
fronds are edible		
Rhodamnia cinerea Jack	Mempoyan	Timber hard though small.
Rourea fulgens Planch. Rourea minor (Gaertn.) Alston		
Rourea rugosa Planch.	Akar prijeh, Akar semilat putih, Keluntut kra, Kelintek nyamok	
Saccharum officinarum L.	Tebu	
Salacia sp.1		
Sauropus androgynus (L.) Merr.	Assin-assin, Chekop	A popular vegetable.

Species name	Venacular Names (Malay)	Uses
Scaevola sericea Vahl	Ambong-ambong	Ornamental, wind screen.
Scaphium macropodum (Miq.) Beumée ex Heyne	Kembang semangkok, S'ilayer	The seed, immersed in water for a night, exudes a gum that contains Bassorin, and used to fill a small cut. The gum mixed with sugar-candy is eaten as an aid for the stomach, and is very pleasant and wholesome.
Schizachyrium brevifolium (Sw.) Nees ex Büse		
Schizaea dichotoma (L.) Sm.	Payung ali, Paku cakar ayam, Misai rimau, Rumput dawai	Used as an aphrodisiac and as a tonic after childbirth.
Scleria lithosperma (L.) Sw.		
Scleria oblata S.T. Blake		
Scleria sumatrensis Retz.	Rumput kumbor, Rumput siamet,	
Sendayan		
Sesuvium portulacastrum (L.) L.	Gelang, Gelang pasir, Saruni, air, Krokot	Edible as inferior vegetable.
Sonneratia griffithii Kurz	Bedada, Pedada, Gadabu, Perepat	
Spermacoce assurgens Ruiz & Pav.		
Spinifex littoreus (Burm.f.) Merr.		Medicinal with anti-microbial properties (dermatophytes).
Staurogyne setigera (Nees) Kuntze	Pokoh checheh merah	Leaves chewed for sore mouth.
Stemona curtisii Hook.f.	Ubi kemili utan xx?	
Stenochlaena palustris (Burm.) Bedd.	Paku miding, Paku akar, Paku naga, Paku ramu	The red young fronds are used as spinach; juice is used to cure fever; and rhizomes to bind baskets and fish traps.
Suregada multiflora (Juss.) Baill. var. multiflora		
Synedrella nodiflora Gaertn.	Rumput babi	
Syzygium aqueum (Burm.f.) Alston	Jambu air	
Syzygium campanulatum Korth.		
Syzygium grande (Wight) Walp.	Jambu ayer laut, Krian batu	
Syzygium zeylanicum (L.) DC.	Nasi-nasi, Merkasih, Glam tikus laut, Beti	Fruit edible, sweet and aromatic.
Tacca palmata Blume	Atjang (Sumatra)	
Tacca pinnatifida J.R. Forst. & G. Forst.	Poko lukeh	Alimentary. A bitter starch called Tahiti. Source of arrowroot.
Taenitis blechnoides (Willd.) Sw.	Paku pijai, Paku balu, Paku pasir	
Tamarindus indica L.		
Teijsmanniodendron coriaceum (C.B. Clarke) Kosterm.	Jali batu, Medang pupei	
Terminalia catappa L.	Ketapang	Almonds edible, very tasty.
Tetracera scandens (L.) Merr.	Ampelas putih, Ampelas hari betina, Ampelas tikus	Leaves used for smoothing wood.
Timonius compressicaulis (Miq.) Boerl.	Mengkudu laut	
Timonius wallichianus (Korth.) Valeton	Menkong	
Tournefortia argentea L.f.		
Tradescantia spathacea Swartz		
Tridax procumbens L.		
Trimezia martinicensis (Jacq.) Herb.		
Tristellateia australasiae A. Rich.		Ornamental, often cultivated in Singapore.
Vaccinium leptanthum Miq.		
Vernonia cinerea (L.) Less.	Rumput sabagi, etc.	
Vigna marina (Burm.) Merr.		
Vitex pinnata L.Leban, Leban hitam,	Leban tandok	Valuable timber
Vitex trifolia L. var. repens	Lagundi, Langgundi	Medicinal
Wollastonia biflora (L.) DC.	Sarunei, Serenah, Sanai laut	
Xanthophyllum vitellinum (Blume) Dietr.	Medang mangas	
Zoysia matrella (L.) Merr.		

ACKNOWLEDGEMENTS

Daven Wu (Editor/Writer)

Jimmy Low (Designer)

Chee Shin Yee for painstakingly and patiently project managing this publication.

The publishing and editorial team at World Scientific Publishing Co. Pte. Ltd.

Cher Chua Lassalvy and Laurent Lassalvy for leading and driving Pulau Tengah's conservation programmes and scientific studies.

The Johor Marine Park and its Director Encik Nizam and Puan Bahrinah binti Bahrim for their collaboration on the Turtle Watch Camp turtle hatchery and the coral reef protection programmes.

The Department of Fisheries for collaborating with us on the turtle hatchery.

The Johor Park Department.

The Johor Forestry Department.

Simon Buckell, the British naturalist, for tenaciously and methodically documenting the bird life of the island, estuary of the Mersing River, and Tokong Bahara.

Dive instructors Kevin Gryce and Anne Jeanneret for their photographs of marine life and taking on the turtle conservation efforts on the island.

Adam Beard and Carmen Toanchina for painstakingly photographing and documenting the marine life of the island and for leading and driving the creation of Turtle Watch Camp.

Dr Bruno Senterre, botanist and scientific collaborator with the University of Brussels for covering every square inch of the island on foot in creating his checklist of Pulau Tengah's flora.

Vincent Chow, President, Malaysian Nature Society, Johor for his contributions.

Dr Louisa Pannampalam, Research Fellow, University of Malaya and Pew Marine fellow 2014 for her scientific contribution on sea mammals and the dugong.

Kee Alfian, Research Fellow, National University of Malaysia for his scientific contribution on marine life.

The University of Malaya (Herbarium).

Forest Research Institute of Malaysia (Herbarium).

Quirijn Coolen an early stage intern at Batu Batu.

Dave Bakewell for his photo contributions.

Eric Frideen for his beautiful photographs.

Rohan Rees, a guest who captured beautiful images of turtles hatching.

Each and every photo contributor who succeeded in capturing the beauty of the island.

And last but not least every single Batu Batu team member and Turtle Watch Camp volunteer who contributed to conservations efforts on and around the island.

A more comprehensive version of the various expert reports can be found online at *www.batubatu.com.my.*

From Left: Abdul Manap bin Abdullah, Kevin Gryce, Bahrinah binti Bahrim, Anne Jeanneret, and Cher Chua Lassalvy

View of the rocks from Batu Batu's restaurant and bar.
(Photo credit: Eric Frideen)

AWARDS

PAM AWARDS 2014

FOR EXCELLENCE IN ARCHITECTURE

PAM AWARDS 2014 · PAM · PERTUBUHAN AKITEK MALAYSIA

PAM AWARD COMMENDATION

Category 9

COMMERCIAL (MIXED DEVELOPMENT & HOSPITALITY)

ISLAND RESORT AT BATU BATU

Architect : Y. ARCHITECTS SDN BHD
Client : BATU BATU RESORT SDN BHD
Builder : JURUSAHA SDN BHD

JURY

Ar Chan Seong Aun, PAM President 2014/2015
Assoc Prof Ar Meor Mohammed Fared, Head of Department of Architecture, Universiti Putra Malaysia
Ar Mike Boon Chee Khian, Arkitek JFN Sdn Bhd
Mr Sathirut Nui Tandanand, ARCASIA President Elect
Ms Kelley Cheng, Editor in Chief, Singapore Architect

CONVENER

Ar Wee Hii Min

Ar Chan Seong Aun
President 2014-2015
PERTUBUHAN AKITEK MALAYSIA

24 May 2014

PAM Awards
2014

FIABCI Malaysian Property
Award 2015
(Resort Category)

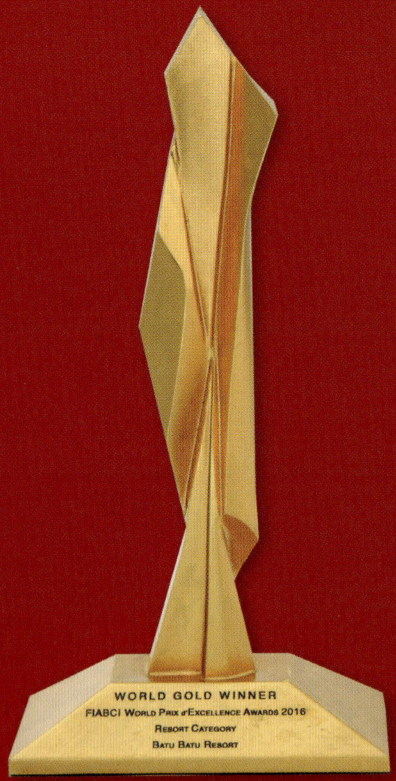

World Gold Winner
FIABCI World Prix d'Excellence
Award 2016
(Resort Category)

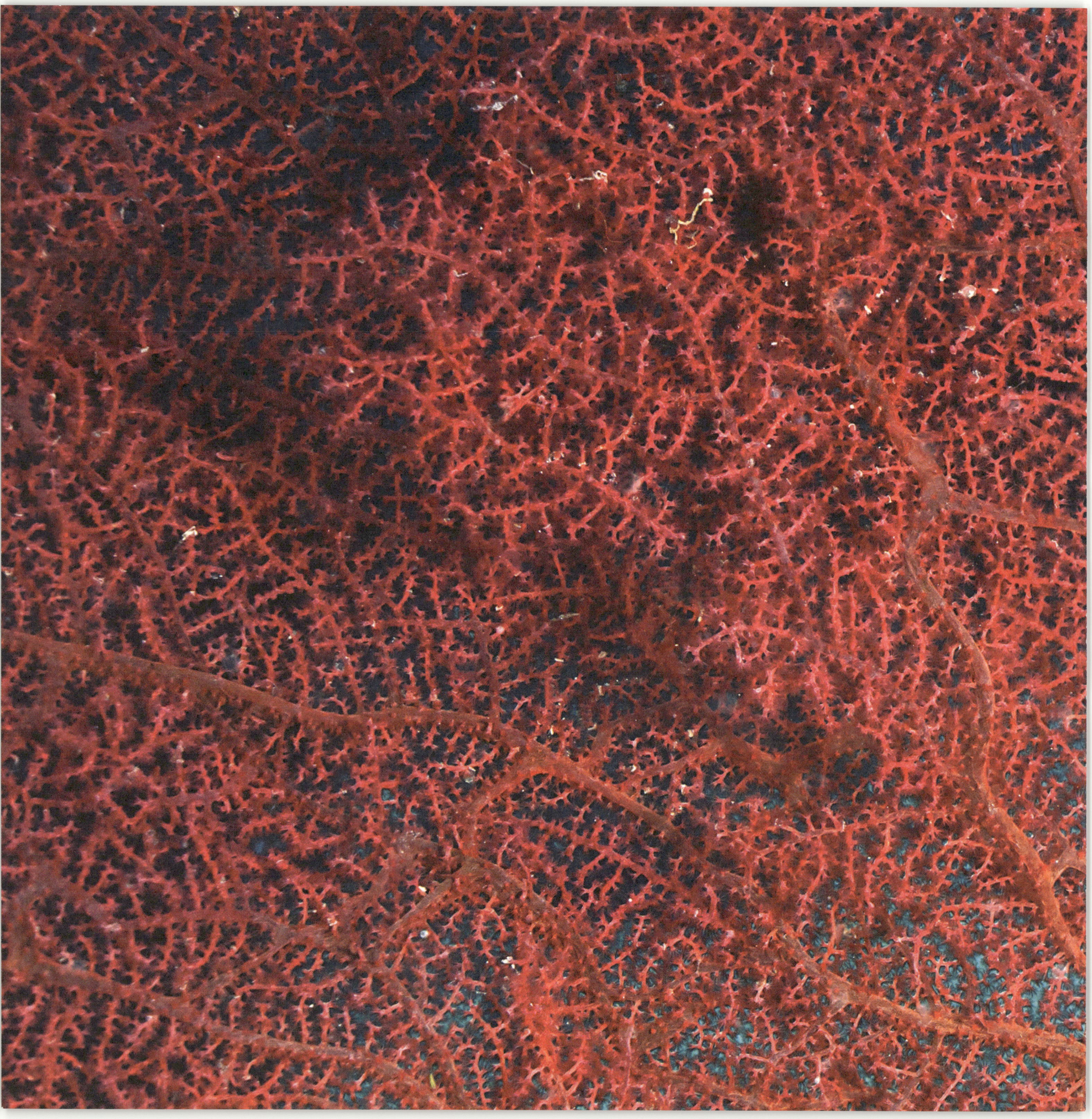